D0046168

The Making of Champions

The Making of Champions

Roots of the Sporting Mind

Gary Lewis

Macmillan
London New York Melbourne Hong Kong

First published 2009 by
MACMILLAN
Houndmills, Basingstoke, Hampshire RG21 6XS and
175 Fifth Avenue, New York, N.Y. 10010
Companies and representatives throughout the world

ISBN-13: 978-0-230-21016-5
ISBN-10: 0-230-21016-3

This book is printed on paper suitable for recycling and made from fully managed and sustained forest sources. Logging, pulping and manufacturing processes are expected to conform to the environmental regulations of the country of origin.

A catalogue record for this book is available from the British Library.

A catalog record for this book is available from the Library of Congress.

10 9 8 7 6 5 4 3 2 1
18 17 16 15 14 13 12 11 10 09

Printed and bound in China

For my parents

contents

figures and box

Figures

Box

preface

A series of debates with a number of sports-minded friends laid the early flagstones for what this book was to become. The question we asked ourselves was just what was it that allowed such marvellous reactions, toughness under pressure, and a hell-bent determination to arise in some individuals, but not in others. What are the roots of the elite sporting mind? Are the deft reflexes of Roger Federer a natural gift, or he is the product of many rigorous years on the training court? Is the ability to soak up the punishment dealt out by his opponents an ability Ricky Hatton possessed from birth, or is it a skill he developed in the tough boxing gyms of Manchester?

We were not the only ones to have this sort of conversation; one of the great pastimes of any armchair athlete will be to dissect the play from the sidelines and prophesise the result of a sporting contest, and why it might go that way. Yet in my initial search to find a more scientific answer to this question, I hit somewhat of a void. The shelves of all too many bookstores were conspicuously devoid of any kind of popular summary or review of just what made the minds of elite athletes as great as they undoubtedly are.

However, what I did start to stumble across in this early quest for answers were a number of books, articles, and documentaries that each told a piece of a wider story, albeit one with many dimensions and shadings. Looking more closely, I began to feel that if the pieces of the jigsaw were there it seemed timely to place them together in a way that one could see the fuller picture. And so, in attempting to draw these threads of the puzzle together, this journey has encompassed various aspects of psychology, cognitive neuroscience, behavioural genetics, and sociology.

In covering such ground my aim was to be as concise as possible in order not to detract from my intention to explore the origins of the elite sporting mind. Nonetheless, I hope those who notice my penchant for veering tangentially in search of a good story or anecdote will not be discouraged by a broader sweep of the brush than a mere reporting of scientific fact. The result, I hope, will be a fascinating dip into the curious world of the elite sporting mind.

In writing this manuscript, and as a first-time author, I have experienced first hand the very great debt any author owes to many individuals for numerous bouts of advice and support. To this end, I must extend particular thanks to a number of people. To Emma Giacon, for being her unpredictable self. She started the ball rolling. To Andy Nicol, for sharing his professional sporting experiences. To Nick Tsatsas, who provided many insightful comments on early drafts of the manuscript. To Pat Collier for the right words at the right time. To Alexandra Dawe, for steering this project through from start to finish. My family, who never let me forget that this was a project worthy of completion. Most of all, a special thanks to Vera Ludwig, whose wonderful support and wise counsel has unquestionably lifted this project.

1
the myth of talent

Amid the tense atmosphere of a Wimbledon Centre Court encounter, the fastest server ever to grace the sport of tennis prepares to unleash yet another firecracker of a delivery. Andy Roddick is a phenomenon in the world of tennis. His service thunderbolts travel at speeds of anything up to 155 miles per hour. Yet despite the fact that the reaction times available to the apparent victim border on the finer side of zero, this 'victim' is in no mood to submit. Instead, he is intent upon, and capable of, unleashing his own brand of punishment right back at the American. The other man is of course the Swiss maestro Roger Federer, lauded as the greatest player of his, and perhaps any other, generation. And where others have failed in this task with such ignominy, today it is Roddick who is humbled. The crowd are suitably dazzled at the display of nimble handedness under such lightning-fast conditions.

Perhaps tennis is not your cup of tea, so instead use your imagination to take yourself to the Olympic athletics stadium and the final of the men's 10,000 metres. Visualise your legs moving faster than they ever have before, and for a far greater period of time. Picture the degree of concentration you will need to control your breathing and your running form; the composure required to deliberate over your tactical race plans amid the heat of battle, as well as to keep an eye on the moves of the opposition. Then imagine that in the heart of this finely poised athletic moment, in the most important race of your life, you are tripped and bundled to the ground in an unfortunate collision with another athlete. Your dreams of victory following years of dedicated training are dashed in mere fractions of a second.

This, in fact, is the very fate that befell the great Finnish runner Lasse Viren during the 1972 Munich Olympics. Yet to the absolute incredulity of the enthralled Munich crowd, he not only pulled himself to his feet and regained a 50-metre deficit but also proceeded to break the heart and soul of the leading pack with his sheer grit and determination. Remarkably, this unflappable Flying Finn took Olympic gold with the fastest time ever run at the distance (one can only imagine the time he may have recorded had he stayed on his feet throughout). The comeback has been rightfully proclaimed as one of the very greatest in sporting history, and places Viren right up there in the pantheon of all time greats.

But maybe a different kind of fighting spirit is more to one's taste; competition of a more carnal persuasion. The 1975 'Thrilla in Manila' – Muhammad Ali versus Joe Frazier – was an animalistic contest, pitting a stinging butterfly against an uncaged, and often unhinged, grizzly bear of a man. The fight was so physically and mentally brutal amid the searing Philippine heat and the barrage of punches delivered by both combatants, that even the winner, Ali, fainted to the canvas shortly after the fight ended. Ali exclaimed that the fight '*was like death. Closest thing to dyin' that I know of*'.[1] Frazier, for his part, never again spoke to his trainer, Eddie Futch, such was his rage at the stoppage of the fight by his own corner at the end of the fourteenth round. Both anecdotes illustrate quite succinctly just how far each man was prepared to go in order to win the fight and the extent to which each man could (and would) suspend physical pain and mental hardship so as to achieve their dreams.

The above vignettes illustrate the powerful role a devout and relentless psychology can play in the outcome of a sporting contest. There is no longer any doubt in the minds of the elite athlete or the layman as to the significance of the mind in sporting competition. As the legendary baseball player Yogi Berra has exclaimed (with his inimitable humour), *'ninety percent of this game is mental... the other half is physical'*.[2] Accordingly, in recent times the mental abilities of the world's elite athletes have become one of the most hotly discussed topics in the pubs and bars across the sporting world.

However, while for the most part we are able to quantify the physical aspects of an elite athlete's performance – such as great strength, flexibility, and stamina – the same is unable to be said of an elite athlete's psychological attributes. We now instinctively know that the psychology of an athlete is important for achieving success at the highest of levels, yet unfortunately, and all too often, we don't know why or indeed how this is so. Is it a matter of superior concentration and focus, a magic 'eye' for the ball or opponent? Or is it simply a higher threshold to hardship that allows some to break through the proverbial pain barrier? The latent nature of the mind does not make this relative obscurity understandable; after all, one cannot readily know that which one cannot see (at least without some clever scientific tools).

Tools for an elite sporting mind

Before we move any further towards an understanding of the origins of an elite sporting mind we must start at the beginning. Indeed, any book that proposes to explain the origins of elite sporting psychology owes the reader an insight into just what an elite sporting mind is before any discussion of how this sporting mind might emerge is proffered. Accordingly, three fundamental components of the elite sporting mind are outlined below, and shall form the backbone of this book's quest.

1. **Elite sporting personality traits** such as high levels of self-confidence and motivation afford one the ability to maximise training sessions, endure extended periods of low success, and so on. Traits such as the dedication that kept rugby star Jonny Wilkinson on the practice-ground long after his teammates are testament to his extraordinary desire to succeed as a penalty kicker, and undoubtedly a great portion of his influence on England's World Cup victory of 2003.
2. **Elite sporting perceptual and decision-making skills** are the abilities that allow one to efficiently process and analyse the visual sporting array needed to make a tactically sound play. How does David Beckham manage to pick the exact right moment to strike his incisive crosses? How do the world's elite boxers slip and slide from an opponent's attack if they are not privy to some kind of inside information? Accurate anticipation and effective decision-making, particularly under stress, are hallmarks of all elite performers.
3. **Mental toughness** refers to the ability to manage levels of stress and anxiety, and to maintain emotional regulation during the heat of competition. The story described at the head of this chapter featuring Lasse Viren is a case in point of great mental toughness and emotional management. Where others might have shown a fit of rage at their poor fortune, and thus totally blown the competitive 'face' all great athletes strive to achieve, Viren simply refused to allow whatever rage he might have felt to overwhelm him.

These three qualities are frequently found in the elite competitors of any sport one may care to mention. Whether one is a golfer, tennis player, basketballer, or boxer, all of these abilities will be crucial to achieving high-level sporting success. They are the basic building blocks from which the foundations of all future successes are founded.

That said, it is of course entirely possible to argue for further aspects of the elite psychological repertoire. One may argue that there are other qualities of equal importance in the chase for success, and that they are notable by their absence. The tactical ability of an athlete is one that may come to mind. Pure tactical knowledge is absent because it does not, to my mind, embody the fundamental differences between the elite athlete and their amateur counterparts. There are many who can dissect a fight, match, or race as well as any world champion. But how many of these pundits manage to utilise this information in a meaningful way out on the sports field amid the heat of battle? Furthermore, there are many highly decorated athletes who display a remarkably low degree of awareness of the wider tactical agenda the elite echelons of sport entail. To this end, I have instead emphasised perceptual and decision-making skill as the real difference between elite and amateur sporting abilities. In other words, it's not what you know; it's how you use it.

Origins of an elite sporting mind – nature or nurture?

We already know that elite athletes possess mental skills superior to those in the chasing pack – any time spent observing a world class performance will remind one of this fact. We also have an idea as to how these skills are grouped and what they represent. But how, and from where, do these great psychological faculties emerge? Is it a God-given gift or a genetic prophecy to be able to sustain the fight to the bitter end, or simply the result of hour-upon-hour spent on the training ground? Is one simply the possessor of a more gifted 'eye' for the ball, or are these skills and reflexes learnt over the hours and years in the gymnasium?

The questions go on, and become increasingly complex. What reaction times do the best motor racers possess; is this the secret of their success? What type of mental training did the Olympic gold medallist do that his nearest rivals overlooked? How do elite athletes focus their minds so completely despite the roar of thousands in the grandstand? And why do some lose their concentration at the moment when it is most needed (along with every other club and park weekend enthusiast)? The list, of course, could go on for some time.

These are not new questions, at least not in the more generalised scope of sporting development, and are most often answered by sporting commentators and enthusiastic lay folk in one of two ways. The first, and perhaps most commonly given answer, contains statements such as '*it was a gift*', or '*he is just a natural competitor*', and '*she always had great eye for the ball*'. Some have even suggested that a form of divine intervention is being enacted, and the post-victory religious-minded gratitude one will often see delivered would suggest that many top athletes would agree with this sentiment; '*I would like to thank God for giving me this opportunity*', and so on. In fact, after the 1970 football World Cup was won by Brazil, a member of the Italian team that lost in the final, defender Tarcisio Burgnich, was quoted as saying, '*I told myself before the game, "he's made of skin and bones just like everyone else", but I was wrong*',[3] in reference to the legendary Pele. High praise indeed. Clearly, Burgnich did not think that Pele was cut from the same cloth as those around him.

The legendary performance of the Kenyan male track team during the 1988 Seoul Olympics is a prime example of feats that commonly attract this form of explanation. The Kenyan runners displayed athletic hegemony at the middle and long distance events not seen before in modern competition. Of the five distance events, the Kenyan men captured four gold medals and were only prevented from a clean sweep by bronze in the 10,000 metres, a fine achievement in its own right. The marathon secured a further silver medal. Perhaps even more astonishing was the observation that the Kenyan team, and indeed many of Kenya's wider community of elite runners, were largely drawn from a tribe residing east of the Kenyan Great Rift Valley. The Kalenjin are a medium-sized tribe of highland dwellers with a population of around 500,000, and have in recent times claimed an estimated 25 per cent of championship medals in world-level competition. The conclusion of many is that they simply must be innately special, or even perhaps blessed by God.

The current scientific advances in genetics (ironically, as we shall see in the next chapter) has added fuel to this argument and it can seem at times to be the stock answer delivery to the question of elite sporting origins; '*it's just in his genes*'. The displays of sporting brilliance that are so captivating often seem to be beyond the realm of the average sportsperson and to some this implies that something radically different must exist within these athletes. How else could one react so quickly, or be so cool under great pressure? Perhaps unsurprisingly, genes, those mysterious entities that we now know play such a formative role to human life, are commonly proffered as the origin of such feats. Clearly, the advocates of the above position stress the importance of nature over the environmental factors commonly associated with the creation of sporting excellence. Explanations such as these have their many advocates. You may know, or indeed be, one of them, and they can be infectious and persuading.

The opposing view often articulates itself with statements such as the elite psychological sporting ability was brought about because of '*the tough/poor background that she grew up in, it made her tough*', or '*he had a really strict coach/father*'. This perspective demonstrates the belief held by many that environmental factors play the fundamental role in one's sporting repertoire, through the medium of one's upbringing. Perhaps a background steeped in challenges moulded a more rugged and able type of competitor. Accordingly, advocates of such a position are usually sceptical of the suggestion that fundamental individual differences, such as elite genetic sporting potential, may be present. Instead, what is argued for is what has been termed the blank-slate position. This is the notion that the human individual is a blank sheet of paper awaiting the writing of the page by the medium of environmental stimuli.

Sport is certainly full of examples of this method of explanation, with many young athletes now placed in specialised training academies so as to develop and hone their talents. The world boxing champion Oscar De La Hoya has spoken of his childhood and adolescent experiences, and his forsaking of nearly all the activities common to one's youth in a drive towards sporting excellence.[4] And this is a common sentiment echoed many times over among the world's sporting elite.

Demystifying the myth

Both explanations – nature and nurture – can be intuitively plausible and appear to be anecdotally appealing much of the time. Yet, as we shall see, there are inherent problems with each. Describing the toughness of Joe Frazier and Muhammad Ali as a talent, or to claim that Roger Federer is the possessor of wonderful perceptual skills would certainly be truthful. Conversely, to suggest that the gruelling training sessions of the women's marathon world record holder, Paula Radcliffe, do not aid the toughening up process in preparation for major races would surely be incorrect. This much is certainly not in dispute. However, what does all of this actually mean as one strives for an understanding of the qualities that underpin great athletic minds, and of their origins?

The answer to that question is surely almost nothing at all. There might be a temptation to assume that words and phrases such as talent, giftedness, and '*he came from a tough background*' offer an explanation of how elite abilities emerge as well as a description. However, this would be to fall for a verbal sleight of hand that can be highly misleading, and is perhaps all too common in everyday conversation. Indeed, to call someone talented says little of what underpins his success by way of superior tactical knowledge, confidence, or mental strength, and so on. This is of course a point that has an importance in areas wider than just the mind. After all, we frequently ascribe the same manner of causation to aspects of physical prowess, such as the punching power of Mike Tyson often referred to as a 'freak of nature', rather than looking for the techniques and biomechanics that underpin such potency. And this is not to mention the many additional abilities we see routinely in the world around us that are explained away in such a manner. One must not forget that the question at hand is where does the talent actually lie?

Similarly, to say that an athlete was simply pushed harder, or experienced greater life hardships fails to speak of what aspect of the teaching and/or the wider background actually facilitated the improvements. Could anyone wake at the crack of dawn, follow the training schedule of champions, or live a toughening ghetto or hothouse training academy life and reap the subsequent reward of elite sporting prowess and success?

Explanations such as those above seem to be a means of accounting for the sum of the parts – as descriptive phrases – but not of what the parts on their own constitute. And this ultimately is surely of greater fascination. What are called for at this point are not simplified and glossy pseudo-science explanations but, rather, analyses that travel to the actual root of the origins of the elite sporting mind. If, as some would argue, nature is at the root of the matter then exactly what gene/s are responsible for the facets of psychological excellence that are so important in winning at the highest levels? This same requirement would also be necessary for advocates of environmental nurture, be it formally or more informally so, as a root to success. Accordingly, what facets of the social world help sculpt and shape the elite athletic mind? One needs to know exactly *what* it is about nature or nurture that mediates greatness, and not simply that they have some arbitrary involvement which can then be used to explain away the greyer areas of our understandings.

Perhaps a more interesting dilemma is the fact that assessing excellence as a simple product of innate origins, or indeed of one's background and nurture, may not simply be meaningless and unhelpful, but in fact downright fallacious. A classic argument often used in defence of the supposed innateness of genius-like qualities displayed by prodigious individuals is that of a 'great eye for the ball'. This alludes to the notion that some precocious individuals see an incoming ball or punch better than their less fortunate counterparts, and thus are privy to a wider range of potential shots and decisions as a result of their innate predisposition. They are believed to have advantageous visual acuity.

This is a form of an argument that has been used in many other domains of expertise. For example, in music, the concept of perfect pitch – the ability to reproduce a musical note without the benefit of any prior reference – is believed to be the pinnacle of innate talent and solely dependent on factors outside the realm of the training ground, or indeed other environmental stimuli. Yet, despite this popular notion of the 'magic eye' existing as the apparently obvious explanation, the fact of the matter is that no such quality has ever been unearthed empirically.[5] This stands in spite of numerous rigorous scientific tests being deployed to try and detect such an effect. Athletes fare no better than the average test subject in tests of visual anatomy and physiology, as well as reaction speeds, outside the domain of their chosen sport. (We shall return to this issue in chapter 4.) Incidentally, the same view is now held of perfect pitch, the remarkable musical phenomenon. And while certain predispositions towards precocity cannot be ruled out, these skills are certainly not the simple product of one's good fortune at being born with the right type of DNA.

The mystique of the remarkable Kenyan athletes also seems to be less impervious once one applies the microscope more carefully to the task in hand. One aspect of particular importance within endurance athletic physiology is the concept of VO_2 max, a measure of the intake of oxygen by an athlete while exercising to their maximum. If the Kalenjin are the recipients of favourable genes for distance running, one might expect higher levels of VO_2 max when both amateur and elite Kenyan runners are compared to their respective non-Kenyan counterparts. However, in a revealing study, the renowned exercise physiologist Bengt Saltin made a striking observation. Saltin noted that a group of young and sedentary/untrained Kenyans showed exactly the same VO_2 max levels as a similarly untrained Danish group of young men.[6] This finding was compounded further by the discovery that elite athletes with extremely high VO_2 max levels were identified in both Kenyan and non-Kenyan populations. The American seven-time Tour De France champion cyclist Lance Armstrong, for example, is reported to have recorded a VO_2 max level of 85 ml/kg/min,[7] an extraordinarily high figure for any person to record, regardless of the individual's place of origin.

So, if VO_2 max is not the secret of success for Kenyan athletes, what could be driving their prominent success? Tim Noakes of the University of Cape Town has suggested a different physiological cause behind the Kenyan success.[8] The notion of running economy often goes unmentioned in sporting assessments but is an important consideration according to Noakes. Good running economy refers to the efficient use of energy resources. The importance of economy is reflected in the observation of disparities in elite athletes' VO_2 max levels, despite similar personal best times. Indeed, it is not unusual for athletes to differ by several VO_2 max level points, yet for the athlete with the lower reading to consistently outperform the athlete possessing the higher reading. The difference is believed to be the degree of one's running economy, the level of efficiency with which the athlete in question utilises his or her resources.

Running economy can encompass several factors such as the increased oxygen uptake ability of skeletal muscle, a biomechanically sound running technique, more efficient breathing techniques, not to mention the psychological qualities that are inextricably bound to one's physical output. In many ways the notion of running economy is not unlike that of a well-insulated house. Keep the heat in and one will avoid the necessity to turn up the dial on the thermostat. Whereas to let the heat waft through uninsulated walls and ceilings is certain to ensure hefty gas or electricity bills.

However, this finding is still inconclusive as a means of demonstrating advantageous genetic predispositions to distance running. This is largely due to the fact that the most effective way to boost running economy is to train hard and fast,[9] something the Kenyan runners are famous for. Indeed, the Kenyan's spartan training regimes, favouring extraordinarily high-intensity runs, with large proportions of hill-work built into their routines, are considered to be the definitive distance training method currently practised (yet by many accounts undersubscribed to by European athletes).

Furthermore, one of the widest reported anecdotes of Kenyan life are the large distances covered daily by children on their way to school or running errands for their family suggesting even the daily life of the average Kenyan is actively involved in defining their athletic precocity. (However, this is not always the case; 800 m-world record holder Wilson Kipketer lived next door to his school and has said he '*walked [there] nice and slow*'.[10]) As such, it is hard to distinguish between genetic predispositions to running economy and the effectiveness of specialised training methods. Ambiguities currently remain as to whether Kenyan genetic predisposition to endurance running is actually on display, or instead if their athletes are the product of extremely effective environmental factors.[11]

This is not to say that genetic advantages do not exist in some shape or form. For one, the lean body shape of so many East African dwellers (and this is not simply an illustration of a McDonald's-free lifestyle – the closest McDonald's to Kenya at the time of writing was in Zimbabwe) undoubtedly aids the budding distance runner (lower body weight means less strain during exercise). However, the simple notion that the Kenyan athletes are possessors of 'miracle' genes simply doesn't hold up to further scrutiny.

Before you jump ship completely, contemplate the other side of the coin, for those who advocate the environmental/nurture argument are similarly on shaky ground. Indeed, those who claim that a tough upbringing, or strict training regimens produce the physical prowess, technical ability, and the mental resolve seen in all elite athletes fail to recognise the fact that for every Mike Tyson, growing up against the backdrop of tough inner city New York, there are surely many others who did not develop the mental tenacity and grit to sustain the spartan life of a world-champion fighter. Many individuals will have grown up through the same neighbourhoods and even families. They may have experienced the very same training regimens, and perhaps in even greater doses than the stars who are used to illustrate the effectiveness of such methods.

What happened to those individuals? Why did they fail in their quest to be toughened like their more celebrated peers? These are not trivial questions.

The reality, when one inspects more closely, is that environmental stress and hardship is as likely to break the resolve of an individual as much as it is to stiffen it. The popular notion that all boxers who rise from difficult backgrounds and/or prison and consequently are mentally hardened is surely wrong. At least as many as those who make it to the big stage, if not many more, will lack self-esteem, confidence, and determination because of their experiences, and these attributes are of course essential ingredients in the armoury of any great sportsman. It is hard not to wonder if Mike Tyson's precocious physical attributes would not have been better harnessed had he enjoyed a more emotionally nourishing childhood. A life spent on the streets and in and out of juvenile offenders institutions could be argued to have created the paradoxically 'fragile-monster'; a man who could brutally knock a man unconscious one day, yet cry on the shoulder of his trainer the next through chronic attacks of self-doubt.[12]

What I hope is becoming clear is that such pseudo-explanations are somewhat insufficient when one wishes to account for the inner workings and origins of elite athletes and their sporting minds (and of course those of their mere mortal counterparts as well). While elite athletes are undoubtedly gifted and talented beings, the tendency to revert to simplified, and even mystical and/or religious forms of explanation can be misleading, and more importantly often untrue. So why have people arrived at such a state of confusion?

Surely the answer lies in the underestimated difficulty of the question. After all, one would not expect the average man on the street to give a coherent account of quantum physics. So, why should one expect such an account of the championship athletic mind when complex scientific areas such as behavioural genetics, developmental psychology, and cognitive neuroscience are so inextricably bound to the question? But sport is such a central part of modern society that an explanation of the skills of its finest protagonists is an understandable urge, even if the answer is lacking in some detail.

An infamous story involving the legendary English explorer and seaman Captain James Cook is a fine example of this apparently human tendency to invoke fallacious answers to explain curious events, which may not be rational or deserving. When his ship, HMS *Resolution*, landed on Hawaiian shores in 1779 the indigenous locals unacquainted with European technological advances were famously said to have immediately deemed Cook to be the incarnation of the revered god Lono.[13] This is a fact many have attributed to the sheer magnitude of Cook's ship. This technological creation was simply beyond the imagination of the Hawaiians of this time. In essence, they used their limited knowledge and understanding of the world and made an inference that Cook must in fact be a God as no other explanation was available to their more scientifically humble civilisation. And Cook was treated accordingly. However, the exalted hospitality Cook initially enjoyed turned sour when the natives realised their error and killed him! Perhaps Cook's fate can be viewed as a (tongue-in-cheek) forewarning for those less developed explanations of championship psychological pedigree.

The road ahead

The time has come for an understanding of greater clarity to emerge, and so my challenge is to explain the origins and development of the three areas of psychological sporting skill mentioned previously; namely, elite sporting personality traits, perceptual and decision-making skills, and mental toughness.

The path we will need to traverse will take in a number of steps along the way towards a qualified answer of that which is of real significance in the making of a champion's mind.

In chapter 2, 'The Developmental Sieve', we will tread into the murky waters of the current debates concerning genes and familial inheritance, and the influences of the environment. As we have seen above already, there are discrepancies with the views that sporting talents can be simply understood as innate in origin, or solely as a product of the environment. We shall see just how the mind, and so the sporting mind, is effectively born through a complex unfolding of innate and external influences working in unison.

Chapter 3, 'Getting Personal', jumps into the broad literature that exists to explain where the key personality traits that shape the development of an elite athlete originate. Such topics as the role of the time spent in the womb, the relationships with one's peers and siblings, and the influence of one's genes comprise an offering of the birth of an elite sporting personality.

The second elite mental quality, perceptual and decision-making skill, is the central topic of chapter 4, 'What You See Is What You Get'. The evidence and truth for a 'magic eye' is explored in greater detail. In doing so, the manner in which elite cricketers are able to cleanly hit a ball moving at high speed, and how premiership footballers are able to decide upon the correct pass amid the heat of battle are discussed. Furthermore, the ways in which one is able to develop such abilities allowing superior visual perception and decision-making in the merest fraction of a second are explored with a nod towards current state-of-the-art sport science.

The third mental requirement of elite athletes, mental toughness, is discussed in chapter 5, 'When the Going Gets Tough, the Tough Get Going'. The psychological skills that elite athletes use to compose themselves under pressure and pump themselves up in times of emotional flatness are described alongside the evidence which suggests that some are simply innately tougher than others.

Chapter 6, 'The Hothouse Kids', takes a step away from the origins of the elite mental characteristics of champions and turns towards the very crucial methods by which elite training regimens aid the development of future sports stars. The essential ingredients of a training schedule, to the importance of rest for a developing athlete, are set against the training memories and anecdotes of those who made the grade.

In chapter 7, 'The Lives of Others', the role of theory is set aside in favour of examples of real life champions, in an attempt to prove that the ideas of scientists are readily apparent in the lives of the top sportsmen. Among the biographical sketches are the formative years of the cricketing legend Ian Botham, the footballing icon Ronaldinho, the tennis champion Andre Agassi, and the golfing superstar Tiger Woods.

In closing, chapter 8, 'Don't Worry Ma (I'm Only Bleeding)', takes a look behind the scenes of elite junior sport considering the ethical issues that have arisen following the rise to professionalism and the problems that vast sums of money mean for a pastime that was once proud and dignified in its amateur status. Taking a full turn from exploring the origins of a champion, the question is posed whether one would truly desire the life of a professional athlete if one were privy to the less-glamorous side of elite sport.

2

the developmental sieve

We find ourselves at a curious point. Perhaps a reading of the previous chapter will have meant that some traditional viewpoints regarding the origins of elite sporting psychological ability will have had their foundations shaken from beneath. It may well have appeared that rather than nature, or indeed nurture, being the responsible party for the elite mental skills seen on display in the arenas of world sport, it is in fact neither. As one shall see, this is certainly *not* the conclusion I hope to leave one with. Nevertheless I do hope that at this moment questions are being raised in one's mind as to the origins of the elite sporting mind.

For those that have already anticipated such a dilemma, perhaps you may be feeling that a straw man has been erected, simply in place to be knocked straight back over again. Indeed, who could surely argue for the case that the lone influence of nature *or* nurture should lie at the root of the matter? One may think that those who possess just a modicum of critical observance will be able to conclude that this simply cannot be a one-horse race. Just a little bit of everything (that is to say, nature *plus* nurture) is surely involved in the creation of a champion's sporting mind, right?

And so, the cat is out of the bag. Nature and nurture are not mutually exclusive entities (as we shall see in more depth shortly). However, despite the fact that both nature and nurture are indeed involved in shaping elite sporting minds, the situation is considerably more complex than one might suspect. The equation we are looking for is not as straightforward as a 'little bit of everything'. This is because nature and nurture are not on a continuum from each other in their influence upon a given trait. It would simply be incorrect to claim that for a given elite sporting trait the influence of nature was *x*%, whereas for nurture it was *y*%.[1] Instead, a process considerably more elegant is at the root of the matter. And so, the time has come for a framework of development to be put forward so as to make headway in this issue.

Delivering such a framework necessitates a dip into the murky waters of genetic science. Although the state of the science can often become impenetrably complex at the best of times, for our purposes it need not be so. In any case, this knowledge is essential as it will be vital for a fuller understanding of the origins of the elite sporting mind (not to mention the origins of the mind in a more general capacity). Let us begin with a brief history of the gene.

When one speaks of genes the door is invariably opened to a world containing intense fascination for many, as well as perhaps a degree of trepidation for others. Indeed, debates over issues such as the desirability of human cloning have claimed major worldwide headlines in recent years. But, however one feels about such matters, we are surely living in the age of the gene. It is hard to pass a single week without some mention of a genetic breakthrough or concern of some sort or another in the daily papers or evening news.

The Nobel prize-winning discovery[2] of the structure of deoxyribonucleic acid (DNA) by James Watson and Francis Crick, the very matter from which our genes are constituted, has allowed insights into the development and construction of the human (and wider animal) body unlike any known before in the history of mankind, and very often in ways that promise great things for humanity. We are increasingly able to decipher the causes of many diseases, the origins of particular traits and characteristics, and even predict certain behavioural predispositions many years before they show any sign of symptomatic activity.

A knock-on effect of such mass public interest in this scientific panacea has been the rush to attribute causation of the great feats of the current crop of elite sporting stars to their genetic heritage.[3] Perhaps this state of affairs is the contemporary interpretation of what might have been attributed to God or a similar metaphysical entity in generations gone by. However, not everyone is aware of what a gene actually is, and far fewer know how genes go about their work, and of what the implications of one's genes might be for an aspiring athlete. Accordingly, let us explore this avenue of development in search of answers to the questions of the previous chapter.

Genes Are Us

Each and every one of us, irrespective of our sporting prowess, possesses 46 chromosomes. These chromosomes are minuscule chemical strands, which are found inside almost every one of the 50 trillion or so cells that make up the human body (to gain some perspective, this number refers to a 50 with 12 zeros behind it). These chromosome strands are organised into 23 pairs, including the pair of XX-female linked chromosomes, or the pair of XY-male linked chromosomes, and are constructed from DNA. Genes are found along these strands rather like a set of beads on a string. In turn, each and every individual gene contains a code sequence that is comprised of four components, or what are termed *bases* – adenine, guanine, cytosine, and thymine. This code is eventually translated into a specific protein by virtue of its unique characteristics (see Figure 2.1).[4]

Knowledge of this sequence of events is crucially important because the living organism is fundamentally constructed of proteins. Indeed, proteins are the building blocks of life in many respects, involved in the processes of our many trillions of cells in almost every way conceivable.

This process of gene-to-protein has led many towards the conclusion that genes are a straightforward blueprint for the construction of the human individual. This perspective alludes to the notion that genes are all and everything where an individual's faculties and personal qualities are concerned, and of course this would by definition include the elite sporting mind. This understanding of genes is probably the kind that leads to the assertion that the precocious young athlete is psychologically savvy and tactically astute in competition because their father was a tough competitor. The same logic is of course valid for physical prowess as well.

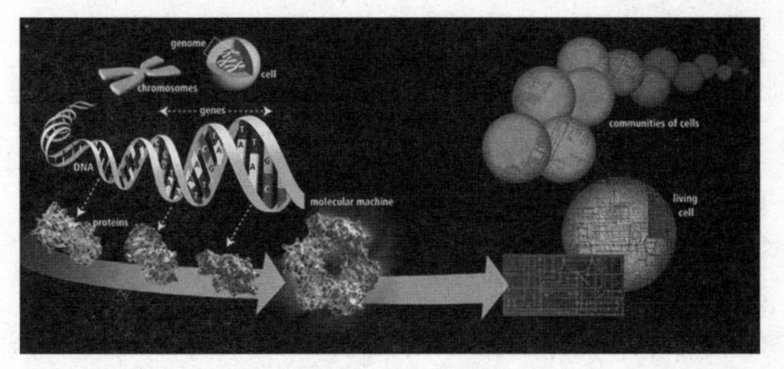

Figure 2.1 From the cell to protein machines. U.S. Department of Energy Genome Programs.

This is not an unfair conclusion to arrive at given the information one will have digested thus far. And a degree of human development is highly fixed from a genetic standpoint, one's blood type or the colour of one's eyes being typical examples. However, in general this conclusion would represent a flaw in any understanding of the role of genes upon development. There is considerably more to the making of an individual than a simple gene-to-protein-to-behaviour pathway.

Blueprint, schmooprint – I'm an individual!

So, what was missing from the paragraphs above that led us astray? The first thing one must recognise is that an individual's genes are, by their very nature, lodged within an environment. And it is this environment upon which they are highly dependent in order to fulfil their coding potential. A crude example of this interdependency between gene and environment can be illustrated with the observation that genes simply cannot function if they are not exposed to environmental necessities, such as nutrients and water. Without appropriate nourishment, living cells will die. With their extinction goes the ability to transfer the basic genetic potential into something more concrete, such as a living, breathing human being able to run four-minute miles or race motorcars at over 200 miles per hour.

The requirements of the environment do not end at the simple provisions of nutrients and water. In certain environments conditions will trigger genetic activity that would not necessarily have otherwise occurred. For example, Siamese cats raised in warm environments develop only a minimal degree of brown fur. These kittens will instead display the cream-coloured coat so synonymous with Siamese cats. This is in stark contrast to cats that are raised in colder climes, whose coats will exhibit far greater degrees of brown fur (this being an evolutionary adaptation – darker furs provide greater warmth as they retain more warming sun rays). It is important to acknowledge that, while both types of fur colour are genetically 'available' in any given Siamese cat, specific environmental factors need to be present for this particular trait to emerge. In this case, the temperature that the kitten is exposed to in the early stages of life is the crucial denominator. This, however, does not mean that certain genes cannot have a greater influence in the emergence of certain characteristics than others, but merely that the environment will also play a pivotal role in any trait that arises.

And so, our genes respond to the environment they are in, switching on and/or off depending on the external stimuli they receive in order to adjust to the changing conditions of the world. And this form of 'push and pull' between one's genes and the environment continues throughout the developmental course. In other words, genes react to circumstances as well as dictate them. The key word of the moment is interdependence. With this observation it becomes clear that the dichotomy between nature and nurture is a false one.

The discovery that far fewer genes exist than was previously thought has, ironically as we shall see shortly, further undermined the notion of genetic blueprints. Human beings are extraordinarily sophisticated animals. We are able to form societies, construct languages, deliver thumping service aces on the Wimbledon Centre Court, and much more besides. It was once thought that we as humans were so complex we simply had to possess an enormous number of genes in order to account for the immense array of feats exhibited in almost every moment of our daily lives. The shock to the scientific community came when the human genome was finally mapped out and it was revealed that in reality we possessed something closer to the order of just 20,000 genes.[5] Previous estimates numbered many times this figure, and the new figure was broadly equivalent to creatures as humble as the sea-dwelling Coral.[6] Muhammad Ali may have been the 'greatest of all time', but he shares at least one common feature with one of the lowliest inhabitants on Earth.

But, how could this be? Were we simply far less complex entities than we had previously allowed ourselves to believe? More worryingly for some, perhaps we were highly constrained in areas of free will and agency by virtue of our apparently restrictive number of genes.

The answer is in fact reasonably simple and thankfully does not mean that either of these fears is substantiated. To begin with, the notion that each gene has a corresponding trait or behaviour needs to be dismissed immediately. Behaviour *y* should not be thought of as a direct manifestation of gene *x*. In fact, what actually happens is that combinations of genes working in unison underpin behavioural traits (alongside the shaping influences of the environment). In other words, extremely rarely will one find just a single gene to be the sole protagonist concerned with the emergence of a characteristic, trait, or behaviour.

So far so good, but what one may well ask is, '*doesn't this simply mean that there are even less available genes to produce the vast complexity of a human*'? If more than one gene is involved with the creation of a characteristic, for example one's height, surely this would mean there are fewer to go round than if just a single gene took care of the job.

Although this is again plausible logic given the information detailed above, this conclusion would be wrong for the reason that genes can, and mostly do, have many more affiliations than just a single group membership. They can help to provide the impetus for a set of circumstances to occur in one particular group action, while helping to abort another set of occurrences in a different group, and later on aiding the initiation of a completely different project.[7] Genes are the epitome of the multitasker, engaged in a number of projects across the developmental lifespan of the organism they reside within. The result is that a small number of genes can exponentially create a huge degree of complexity by virtue of their overlapping and intertwined workings with both the environment and their biological colleagues.[8]

What this shows us is that genes do *not* simply equal a blueprint in the development of complex characteristics, as is so often and commonly perceived, and of course neither does the environment. Rather, they are bedfellows, defined by the other, and in fact in existence only because of the other. As the evolutionary biologist Richard Dawkins has noted, one cannot infer a complex human characteristic from a genetic blueprint any more than one can claim a cake crumb was derived from a specific ingredient used in the baking process.[9] Instead, the characteristic will have emerged as the product of collaboration, and a rich and highly complex one at that, between a spread of colluding factors. And this will be as true for Lennox Lewis, Pete Sampras, Paula Radcliffe, and Tiger Woods, as it will be for you and me.

The development of personality

So, interdependency is crucial in the process of all development. But what does this actually mean when one speaks of an elite sporting characteristic, such as the three fundamental psychological traits outlined in chapter 1? How does this process work? The next chapter, 'Getting Personal', is devoted to the influences that shape the first characteristic – an elite sporting personality – and so it seems appropriate at this stage to illustrate this process through the lens of personality development.

Traditionally, personality research has focused on the measurement of traits. Psychometric tests are well known to near all in this day and age through their common usage in corporate and/or educational settings to identify the strengths and weaknesses of potential employees or students. This practice is also increasingly finding its way into the sporting domain and being used by coaches to distinguish those with desired qualities from those without.[10]

Yet psychometric testing, and indeed much of the literature on personality for better or worse, is essentially a descriptive tool. They allow the tester to understand some of the characteristics of a given individual's personality. However, the issue in question here is what *makes* a sporting personality the way it is and not merely *what* it consists of. Insights yielded by the genetic and psychological sciences, some of which were alluded to in the outline of genetics above, have encouraged a new insight into the collaborative development of the human mind between both nature and nurture. The upshot has been that models of development have emerged, which allow one to begin to understand just how an elite sporting personality might be formed.

The first aspect of this process necessitates a reminder – genes code *only* for proteins, and *not* for any form of complex behaviour *per se*. Accordingly, there is no such thing as a gene for confidence, or determination, or indeed any other complex psychological characteristic that might be desirable to an athlete, despite what the popular press may claim to the contrary at times. However, each one of us *does* possess a unique kind of neurophysiology from very early in the development of life, by virtue of one's genetic heritage. This unique activity, in turn, will direct the individual in question towards basic behavioural tendencies. One example of such basic tendencies can be illustrated when something unexpected happens in the visual array of a young child, such as the appearance of a new and unknown face. The novel event will trigger activity in the brain, in part as a defence mechanism against potential danger. For the infant with a high threshold (or lower levels of activity) for novel occurrences it is more likely that he will be comfortable with the new situation and perhaps even investigate the novelty. However, a lower threshold in another infant may cause them to cry and/or protest at the unexpected event.

The same kind of process could be attributed to other basic forms of behaviour, such as an infant's level of activity. Some infants are relaxed and sedate, whereas others often appear to have somewhere else to be, such is their relentless movement. Again, one's neural makeup will play a fundamental part in such behaviour. And the same will be true for all individuals whether they might later become the captain of a winning World Cup football team, or simply the patriotic fan observing keenly from the terraces. One should not forget that at this point the level of behaviour described is very much below that of conscious processing and is most reflective of a young infant not yet fully able to make a complete grasp of the world around them. Yet, the fundamental building blocks of life are now present from which greater complexity will shortly emerge.

We have reached a point where the causes of some basic differences between individuals can be seen, albeit in a crude manner and very early on in the development course. But still, this is hardly what one would call a personality. After all, a personality is a collective of beliefs, values, and modes of behaviour that are frequently complex, and not simply basic tendencies that incline one to investigate novel stimuli, or other similarly crude behaviours. Accordingly, how do we bridge the gap between the basic tendencies influenced by the neurophysiology in the brain of each individual, and the complex and unique personality traits we each possess and see around us on a daily basis?

This explanatory gap can be bridged by, acknowledging that humans live in a world of considerably greater complexity than all other of nature's creatures, consequently we have evolved strategies and abilities to handle this challenge. Accordingly, in order to cope with the world around us, and to make sense of the enormous array of stimuli that bombards us at virtually every waking moment, we categorise the information into what are often referred to as '*meaning-systems*'. A meaning-system is simply a set of explanations, values, and/or beliefs about events, people (including oneself), and objects, providing a framework with which to understand the world and our experiences. In essence, they are a kind of database that we use to reference our judgements of future action, as well as to make appraisals of past events. For example, one might view upper-class individuals as aloof or snobbish when in the company of someone of (perceived) lower social standing. Accordingly, one might explain an act of rudeness from such persons on the basis that their social standing prompts such behaviour whether or not this is accurate to the truth. In short, we understand the events of the world by using one piece of information to benchmark further insights.

We are edging closer to the complete model. However, we need one final piece in the jigsaw before viewing the completed picture. The concept of meaning-systems is a well documented and popular way of explaining many of the inner-processes we utilise in order to navigate ourselves in the social world. But these meaning-systems are not simply arbitrary representations. They need to be consistent with the content in the rest of one's other meaning-systems, or else the way in which one understands the world is liable to become contradictory and incoherent. One will experience what is termed *cognitive dissonance*. Let me explain the meaning of this term by way of a short story.

The year 1844 was set to be a major event in the faith of the Millerites, a Christian religious denomination in the US. According to their interpretation of the biblical scriptures Jesus Christ was to be received upon earth on the 22nd of October, and the Millerites' waited in understandably fevered anticipation for his arrival. However, to the enormous disappointment of the Millerite movement, Jesus Christ did not make an appearance that day, nor indeed the next, or even the subsequent weeks, months, and years.

The upshot of this absence has been termed the Great Disappointment of 1844, and is of particular interest in this discussion of personality because it examples the notion of cognitive dissonance with some elegance. The Millerites were faced with a dilemma with some considerable implications for their faith. Jesus Christ had not arrived, and accordingly, certain beliefs held in the meaning-systems of these people now contained an inconsistency of rather epic proportions. It was not, of course, possible to have trust in both the biblical scriptures, and the evidence (or indeed lack thereof) of one's own eyes. Certain aspects of the beliefs of the Millerites were in cognitive dissonance and something would have to give. In order to bring the system of meanings back into a more coherent balance, modification in one or more belief/s was necessary. And in fact this is exactly what was observed. Some of the Millerites did this by renouncing their belief in the scriptures, albeit after a further period of waiting for the return of the Messiah. Others, perhaps those more heavily invested in the movement, simply challenged the interpretation of the scriptures and instead erased this aspect of their meaning-system.[11]

We now possess the necessary ingredients to provide a developmental explanation of personality traits. As we have seen, each individual has a unique neurochemistry influencing basic behaviour tendencies. Each of us is also coupled with a faculty, and indeed a necessity, to develop meaning-systems, alongside the need to maintain some level of internal consistency. From these observations it is possible (albeit in a crude manner) to understand where a personality is derived from. For example, if a child possesses a low threshold of novelty seeking, the outside world (family members, friends) will, in time, label the child in a number of ways. This might perhaps begin with 'shy', before leading onto more complex assessments. This will allow these individuals to incorporate the child into their own meaning-systems. The child will also develop its own internal labels, which will require consistency so as to avoid the discomfort of cognitive dissonance. It will be hard for an instinctively fearful child to consistently convince themselves they are a thrill-seeker, particularly when others also affirm that they are not, and so the gradual process of moulding a personality progresses. One's basic tendencies, exposed to the vastness of wider society and culture, will eventually yield a set of traits, beliefs, and values, which are highly unique,[12] as Figure 2.2 illustrates.

And so, as one can see with the example of personality, the cascade of human development is highly complex and littered with moments of importance and influence in a variety of guises. Figure 2.3 illustrates this complex development with the image of a 'rough-terrain' presenting a number of different pathways

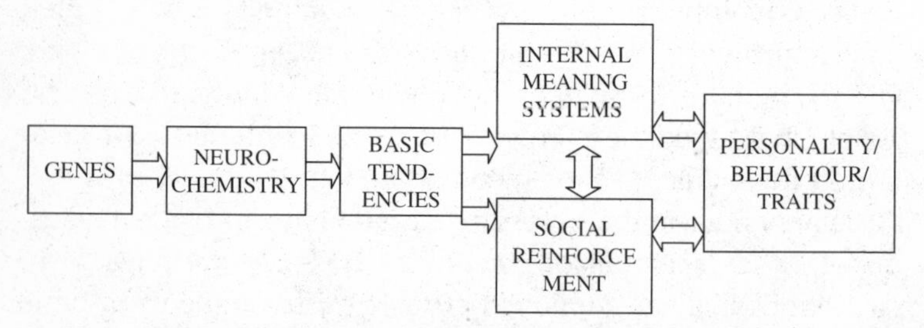

Figure 2.2 A developmental model of personality.

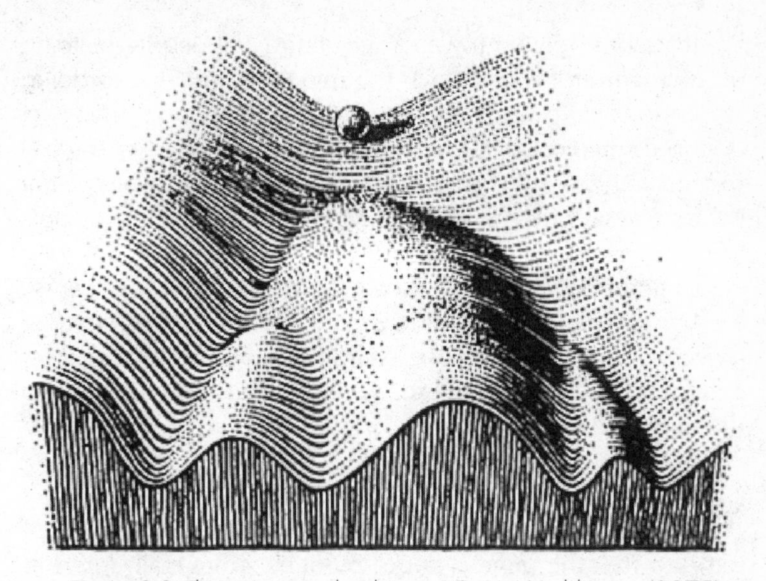

Figure 2.3 The epigenetic landscape. C. H. Waddington (1975) *The evolution of an evolutionist*, Edinburgh University Press.

for a rolling ball (or human life) to travel along, in what is known as an Epigenetic Landscape (time = epi, and change/creation = genesis, as derived from the Greek, referring to the process of change over time). Each potential pathway is dependent, in part, upon the starting conditions, and of course of what occurs following those starting conditions. Importantly, this principle is not exclusive to personality development. Rather, the rise to complexity for all sophisticated traits can be viewed in such a manner.

The ball at the peak of the rough-terrain can be likened to the individual in question, with the bumps along the track alluding to the complex genetic-environmental interaction one will experience throughout a lifetime. If one imagines the ball (or person) being released from the top of the slope, a hypothetical developmental pathway can be viewed. Perhaps the ball will roll down the first slope veering towards the centre, yet despite building up momentum fail to breach the middle hump, and in doing so canon back towards the left field area. Of course, the exact opposite could occur, wherein the insufficient speed of the ball to breach the middle hump in the first example may be greater in another set of developments, and as a result the individual will be drawn towards a different life trajectory, perhaps with very different outcomes. The epigenetic landscape is of course a rather crude representation of the sporting developmental trajectory, and in reality one will be exposed to a far greater panorama of bumps in the road. Yet the illustration provides a useful way to visualise such a complex process.

The Matthew effect

It is now that this genetic background becomes really interesting, for this interdependence of nature and nurture emphasises two fundamental points. Firstly, a particular combination of genes and environment may have a very different outcome in the development of an individual, even if the initial differences were only discreet between persons. Secondly, what might be a strong influence upon an individual's development at one point in time may well be wholly inconsequential at a later moment in time.

A hypothetical example illustrating these points would be a young athlete attempting to resist the temptation of their peers to go and drink with them in the pub the night before a big tournament. Whether or not the young athlete succumbs to the temptations will likely depend upon their set of values and beliefs regarding their sporting goals and also their personal moral code at that given moment. A focused and dedicated young athlete will be unlikely to jeopardise their chances of success. However, this very same individual a year or two down the line, perhaps disillusioned with the strict discipline required for elite success, might have developed a different set of values within their meaning-system and thus eagerly grab the opportunity to rebel. With such a notion in mind the Matthew effect, a historically rich term, becomes of some relevance to the origins of the elite sporting mind.

The Matthew effect refers to a biblical passage in the Gospel of Matthew stating, '*for unto every one that hath shall be given, and he shall have abundance: but from him that hath not shall be taken away even that which he hath*',[13] essentially alluding to the rather unfortunate concept (for the poor at least) of '*the rich get richer, and the poor get poorer*'. What this passage illustrates is the notion that the opportunities exercised by those with wealth and influence tend to compound exponentially, typically meaning that the mere ownership of wealth can be enough to create additional wealth (interest gains on savings, opportunities to invest in financially rewarding stocks and shares, and so on). Conversely, those without wealth or influence tend to struggle to engage in matters that might be financially rewarding, and as a result may suffer in relation to their wealthier counterparts.

In educational terms it has been argued that the age a child learns to read is significant because it leads to challenges, such as more difficult books, that promote cognitive-development. The small advantages gained through an early and advanced development gradually apply a knock-on effect throughout infanthood, childhood, and adolescence. In time the child may find himself in a more promising position compared to a child who has not developed reading skills at a similar age.[14] Similarly, if one is exposed to more challenging intellectual ideas at a younger age than one's peers, there is likelihood that one may tackle more sophisticated endeavours with greater success from a younger age, and this process may continue until similarly bestowed children may display very different skills in a given domain.

What the Matthew effect also alerts one to is that while one may teach a particular technique, say the heavily spun David Beckham or Roberto Carlos free kick, the ability to learn that technique at any given moment will depend upon the developmental trajectory of the child. While one eight-year-old will grasp the cognitive challenges of the technique of kicking with spin, and demonstrate the mental resilience to endure the learning process when difficulties present, another might not yet be able to do so. The ability to successfully learn the kicking technique will in its own way compound the issue. The child might make the local football team because of their improved shooting abilities or receive confidence boosting praise from the coach, which in turn will lay the flagstones for future positive advancements.

One charming anecdotal example of these 'quirks in the road' can be seen in the life of the great heavyweight-boxing champion Muhammad Ali. When Ali, then known as Cassius Clay, was just a tender (and, reputedly, very slender) 12-year-old, a thief stole his treasured and brand new bicycle. The bicycle had been a much-loved birthday gift to Ali from his family and the theft invoked his rage so much so that he threatened, in what was to become his trademark style, to 'whup' the thief if ever he were able to find him. Unfortunately for Ali (although perhaps not for the thief) the identity of the villain remained a mystery, and so Ali was denied his chance for retribution. However, the local policeman who heard Clay make this tearful boast reportedly listened to the young Ali's threats and protestations for a short while before asking young Cassius if he really knew how to fight quite as well as he could talk. Clearly, the early signs of he who was to become, the legendary 'Louisville Lip', were present early on. And upon hearing that Clay was in actual fact a complete fighting novice the policeman invited him to train in the amateur boxing gym he ran for the boys in town so as to teach him how to 'walk the walk' as well as 'talk the talk', having taken a shine to the young lad's boisterous yet charming ways.[15]

The rest of the story is now of course etched in the annals of both sporting and popular history. But if it were not for Ali's early swagger perhaps he would not have been drawn to the attention of his first trainer, and subsequently never have been exposed to the sport of boxing that was to make his name. The very nature of the young Ali could be argued to have drawn him towards the path he set off, however crudely one wishes to pin his later success and fighting fortune upon this curious moment in history.

Perhaps a more poignant tale is that of the youngest player ever to break into the tennis world's top ten rankings (at the tender age of 17), the United States' Aaron Krickstein. Krickstein was by all accounts a shy and introverted teen, lacking in confidence and self-esteem despite being one of the great success stories in US junior tennis history. His coach at the time, the legendary Nick Bollettieri (responsible for the careers of such greats as Andre Agassi, Monica Seles, and Jim Courier), has described the difficulty in instilling Krickstein with an inner confidence and getting him '*out of his shell*'.[16] Ironically, as Bollettieri says, a relatively arbitrary book with a set of motivational quotes given to Krickstein to read by his coaches seemed to spur the breakthrough, something the more formalised coaching had failed to do up until that point, or had it?

We are all more or less receptive to a new idea depending on the stage of life we are in at a given moment. Krickstein may well have just been ready for the aphorisms he discovered (whatever they might have been). In other words, had he developed to the point where he could take the new and positive concepts on board in a positive manner? The answer would (at least anecdotally) appear to be in the affirmative and this experience is surely one that every one of us has observed in our own lives. One is simply not ready to learn until one is *ready* to learn. For anyone who has had the fortune to work with young athletes, this is only too apparent. And so this seemingly obvious statement is perhaps considerably more profound in nature than it is often given credit for. Perhaps the most interesting question with regard to the development of an elite athlete is whether or not that awakening could have emerged earlier.

These are, of course, just anecdotal examples of the issues that may arise in a developmental setting, but they illustrate the kind of moments that are typical in all children's learning, and they are certainly many in number and frequent by occurrence. The message that should be received at this point is that individual experiences are unique, and that each and every individual's experiences will send them off upon a trajectory that in turn will spark new experiences, which in turn will signal new trajectories of their own. A coaching session delivered to a dozen children will impart a very different message to each of the dozen, depending on their own personal level of development at the time in question. The upshot of this developmental process is that each and every individual is blessed with a kind of magical wizardry. But, rather than the magic lying within an ill-defined mysterious entity, such as the 'great eye for the ball' or 'mental toughness', the magic instead lies in the uniqueness of the development each individual is subject to over the course of their formative years.

The road ahead

The challenge that lies ahead at this point is to delineate this complex developmental process and extract the areas of key importance in the making of champions. The obstacle in our path lies in the knowledge that separating the genetic and environmental influence upon an individual is next to impossible. The two are dependent upon the other for any individual, and do not provide discreet and separable portions of influence. They are necessarily intertwined, somewhat like the two sides of a single sheet of paper.

Some qualification is needed here for this notion is not strictly correct, as the next chapter will show. There are ways in which to estimate the importance of one's genes in the outcome of certain complex traits; studies of identical twins raised together and identical twins raised apart allow scientists to infer a proportion of variance in certain abilities to the genes. However, this is not a feasible possibility for measuring the role of a particular teaching method or environmental feature on an individual's later success. One might measure these environmental aspects for evidence of importance through various other means, but they cannot be quantified *per se*.

This might be a very great disappointment for anyone who has made it thus far and was hoping for an exact guide to the origins of an elite sporting psychology. Unfortunately, such knowledge is simply beyond the scope of science in the current age. Nonetheless, although the exact equation that defines the elite sporting mind is not available, great steps have been made in recent times to illuminate the workings of the elite sporting mind that are sure to prove curious enough in their own right to those searching for insight into the phenomenon. And these influences can still be thought of as variables in an equation of sorts; indeed, those athletes with a healthy dose of a range of the favourable influences to be outlined shortly will essentially find themselves on the more favourable tracks to obtaining an elite sporting mind.

This situation also emphasises the role of the coach as 'artist', someone who represents more than a mere box ticker of an athlete's progress. As one cannot hope to discover a wholly prescriptive scientific model of coaching, the need for instinct and experience to guide one's way through the development of a young athlete is crucial. A good coach will certainly know of the fundamental environmental inputs that all athletes must receive to develop in good time. A great coach also knows the right moments to apply these influences in greater or lesser doses. Whereas the time is often right for the infamous 'hairdryer' treatment, *à la* Manchester United's Sir Alex Ferguson, there are also times when subtler approaches are the best route to influencing an athlete to achieving better feats. If one recalls the notion of the Matthew effect, one particular approach may be fruitful in certain situations and utterly useless in others, even within a single athlete. The legendary boxing trainer Freddie Roach has shown his awareness of this process when speaking of world champion Manny Pacquiao – *'You can't be a dictator with him… You tell Manny to do something, he'll go the other way. It's a constant negotiation, but there's a trust we have and an understanding of each other'*.[17]

We have reached the goal of the chapter; namely, to provide a framework of development with which to understand the complexities of the genesis of life and of the origins of the elite sporting mind. While the answer proffered thus far has perhaps raised as many questions as it has answered, nonetheless, the traditional view of elite sporting origins can be firmly laid to rest. Hence, the task that now lies before us is to explore the curiosities that conspire to mould the rough-terrain of life into a unique human individual, and to impart the key personality traits that underpin sporting excellence. Personality took a central role in the illustration of this rough terrain and so it is fitting that this will be the next stop in this quest to provide a semblance of the answers posed in the opening chapter.

3
getting personal

The last chapter argued that human abilities do not emerge from the influence of nature or nurture alone. These two factors were seen to be utterly interdependent in the developmental cascade of life. This has meant that directly quantifying the influence of any particular aspect of an individual's development is just not possible. And so, understanding how much of an elite athlete's abilities are because of factor *x* is beyond modern science.

Nonetheless, the numerous influences upon any complex trait – such as an elite sporting psychology – are becoming better understood. And even if exact measures of influence are not possible, illuminating some of the likely influences is important in its own right. Accordingly, the job ahead of us now is to explore the factors – both innate and environmental – that conspire to sculpt the mind of an elite athlete. In particular, the task of this chapter will be to tackle the origins of the elite sporting personality. However, before we continue to the influences upon the elite athlete's personality, a fundamental question must be posed. What is an elite sporting personality?

At first this might seem like a strange question. After all, how could there possibly be a generic personality-type for all athletes, in any number of diverse sporting pursuits? It is clear that even within a single sport there are enormous differences between the personalities of participants. Those who recall the five-time Wimbledon champion Bjorn Borg exchange backhands in anger with the brash and hot-tempered American John McEnroe will know how very different the two men were. On the court McEnroe was overt, angry, impulsive, and petulant, whereas Borg was emotionless, calm, and thoroughly methodical.

To further complicate the issue, there are sports where the various playing positions require very different mindsets to others from within the same sport. England goalkeeper David James has commented that '*the best teams have a combination of psychological make-ups – your obsessives in the back line, and one or two in midfield, who increase your chances of winning through their hard work and repeated practice. Then you have the flair players who display flashes of genius, of brilliance and unpredictability*'.[1] This seems to be reasonable logic. After all, it is hard to imagine footballers such as the Brazilian superstar Ronaldinho playing at right back. His opportunistic nature would surely be stifled in such a deep and mostly conservative position.

But there is a problem with this appraisal of the sporting personality. Referring back to the Borg–McEnroe dichotomy, it is only too clear that each competed very differently on the tennis court. Yet what of their similarities? Both men were certainly highly confident of their abilities and displayed extraordinary tenacity to fight their corner to the very end, a quality both men demonstrated on many occasions. Could it be that under the surface a very similar set of qualities exist, obscured by those more overt displays of character?

The short answer to this question seems to be yes. Elite athletes from many different sports do indeed share a core set of personality traits that are believed to underpin their tremendous competitive success. This is not to say that individuals within and across sport are all alike – uniqueness is only too apparent. However, when the gloss is removed, striking similarities in the makeup of elite athletes' personality appear time and again.[2]

Elite athletes are more highly disciplined, more conscientious, less introverted, and lower in anxiety than the normal population. They also possess, on average, greater emotional stability, self-reliance, and self-confidence than their non-athlete peers.

This outline of the elite sporting personality prompts two important issues. The first is that while these qualities are present in most athletes, the way they are expressed can be very different. Self-confidence is perhaps most evident when someone 'trash-talks'. John McEnroe was not beyond telling an opponent just how bad he was going to '*whup their butts*', when he was in his most boisterous of moods. But just because Borg was not taken to such behaviour does not mean he was not similarly assured of his abilities. The fundamental point is that the nuances of personality are going to be unique. But the deeper-rooted traits will often be the same between elite performers.

The second issue is that an elite set of personality traits is not going to guarantee any given athlete a shot at the big time. One recent study illustrates this point having followed a series of elite junior Australian Rules Footballers in their attempts to make the transition to the professional game. At first the personality traits of the young athletes did not predict their likelihood to make it in the senior game. However, once the differences in physical ability were controlled for, personality became a strong predictor of professional success.[3] The moral of this tale is that personality, like most other components of the elite athlete, is not the be all and end all. However, all else being equal, a favourable set of traits will probably take one further than another athlete who is lacking in some of the important characteristics.

We now have a sense of just what personality traits are fundamental in propelling an athlete to the highest plateaus of sport. Thus, what is now required are the variables in the equation that can instil these kind of personality traits. In short, what kinds of factors are crucial in shaping the personality traits that can propel a young hopeful towards the bright lights? In the previous chapter, one's basic tendencies were shown to be an important platform from which the sculpting of one's more complex personality traits emerge. But what are these basic tendencies and what defines their nature?

Back to basics

If one were to ask a thousand people to answer a number of questions about their personality, two things will tend to happen. Firstly, there will be considerable variance in the population; people are not all the same and often provide very different sets of answers. However, the observation has been made that a small set of factors seems to underpin what at first appears to be a very large set of characteristics. This is demonstrated by statistical tests that illustrate the shared variance among the population for a given set of characteristics. That is to say, rather than characteristics such as (a) enjoys foreign travel, or (b) enjoying 'thriller' movies, existing independently from each other, they both tend to be high (or low) in any given individual. Accordingly, psychologists have been able to identify a set of five factors that appear to underpin most other characteristics.[4] These five basic traits have been referred to as the five factor model and consist of – openness, conscientiousness, extraversion, agreeableness, and neuroticism.

The factor of openness refers to the degree with which an individual embraces new situations and ideas, the awareness and openness to his feelings and emotions, and the values held by the individual. The conscientiousness factor encapsulates qualities such as an individual's self-discipline and degree of orderliness. Extraversion is the factor describing an individual's degree of traits such as assertiveness, excitement seeking, and gregariousness. Importantly, extraversion in this context does not imply sociability, as Carl Jung thought, although an extravert may be compelled towards social interactions. The factor of agreeableness refers to attributes such as the degree of an individual's altruistic tendencies and the compliance to the demands of others. Finally, the neuroticism factor involves an individual's degree of anxiety levels, self-consciousness, vulnerability, and emotional resiliency (although this is not necessarily, despite being commonly assumed, a pathological term in this context).

There is evidence that elite athletes are lower in neuroticism and higher in extraversion than the rest of the population.[5] This makes intuitive sense, as those who are more highly drawn to the excitement of victory and are less shaken by a loss are more likely to possess the drive to continue on their quest to elite status. It is also self-evident that high levels of conscientiousness will be crucial to an aspiring athlete. The predisposition to complete one's training to the best of one's ability is paramount to making great improvements in one's game. The Olympic champion sprinter, Michael Johnson, has supported this sentiment – '*I have a training log, a schedule of when I will run, how far I will run, when I will lift weights. If it says to run twelve 200-metres intervals with 2 minute's rest, that is exactly what I do. I don't blur the edges of something that important*'.[6]

The basic tendencies of agreeableness and openness are perhaps less obviously involved. Certainly, an elite athlete will require a degree of ambivalence towards their competitors – such an individual might score low for agreeableness. Yet, they will also need to cooperate with their coaches and/or teammates. Openness suggests an athlete will be keen to learn new skills, a useful characteristic. We saw some traits of elite athletes above, but now let us view them in line with the five factor model. Yet how much openness one needs to make the most out of a well-meaning coach is debatable at this moment.

We now have an understanding of the basic tendencies from which one's complex personality traits emerge. However, what defines the nature of these basic tendencies? As one will recall from the previous chapter, genes provide an indirect pathway to an individual's personality traits by virtue of the unique neurophysiology that defines one's basic tendencies. As the renowned personality researcher Marvin Zuckerman has noted, '*We are not born as extraverts, neurotics, impulsive sensation seekers, or antisocial personalities... we are born with differences in reactivities of brain structures and levels of regulators*'.[7] What Zuckerman refers to are the differences that exist across the population in how our brains respond to environmental stimuli, including of course other people. For the purposes of the elite sporting personality we shall explore three of these basic tendencies – extraversion, neuroticism, and conscientiousness – as these three tendencies are clearly central in the makeup of an elite athlete.

Let's begin with the extravert – that individual who engages with the world and feels a sense of 'aliveness' that not all do. Indeed, whereas one individual (the extravert) will be overcome with positive emotion at the thought of, say, winning the lottery, another may be pleased but may be somewhat indifferent to their good fortune (the introvert). We now know that extraverts are, in part, the way they are because they possess a higher level of responsiveness in brain areas that comprise what is termed the *reward system*. That is to say, areas of the extravert's brain that signal the potential of a reward (the chance to win a medal, and so on) – such as the ventral tegmentum and the nucleus accumbens – are known to be more responsive than those same areas in the brain of an introvert. Those with a less responsive reward system are more likely to be introverted. Things just don't seem as worthwhile for these individuals as they might for the extravert who just can't contain himself at the thought of a possible gold medal. However, this does not mean that introverts are necessarily unhappy. A low level of responsiveness to possible reward is not the same as feeling blue. As with the reluctant lottery winner in the example above, someone veering towards introversion is more likely to be characterised by a degree of indifference, than by fear or sadness. As psychologist Daniel Nettle has remarked, '*the introvert is, in a way, aloof from the rewards of the world, which gives him tremendous strength and independence from them*'.[8]

While certain brain areas respond strongly to potential rewards, there are other brain areas – such as the amygdala – that process the negative emotions people experience. And as you may have already guessed, those who are more neurotic will be likely to have greater responsiveness in these brain areas controlling fear responses. Intriguing studies have been conducted in recent years that illustrate the different levels of responsiveness that exist in these areas. Indeed, when participants undergo a brain scan there is great variation in the response of the amygdala to fearful faces. Some people are simply more negatively aroused by troubling events than others – such individuals typically score higher on levels of neuroticism.[9]

Let us move onto the next trait. Conscientiousness, if one recalls, is the characteristic where an individual pursues a goal with particular rigour. And while this might at first sound like a feature of a highly responsive reward system, there is an important distinction. Indeed, for whereas one might be compelled to partake in a given activity, the conscientious individual will be able to inhibit his urges if the activity is counterproductive to his goals. A young athlete might feel the compulsion to skip training for the day and go to the cinema with his friends. And his level of conscientiousness will dictate what course of action he takes.

Indeed, high levels of conscientiousness will mean that one is able to inhibit the signals from one's reward system. The frontal lobes of the brain – the so-called executive region of the brain – have been implicated with the faculty of conscientiousness. Numerous studies have shown this link, perhaps the most well-known being those that have used the Iowa gambling task (IGT).[10] In the IGT, participants are asked to draw cards from one of four decks, with each pick typically yielding a reward, and occasionally a loss. The aim of the task is to pick cards that maximise one's overall gains. However, unbeknownst to the participants, two of the decks are 'good' decks, while two are 'bad'. The good decks are low earners, but even with the occasional losses they still leave one in profit. The bad decks, however, offer far greater rewards but the losses incurred mean that one is liable to come out of the task in debt. People who have suffered brain damage in the orbitofrontal cortex (the region of the brain directly behind the eyes) perform poorly on the IGT and continue to pick the bad deck even though the losses gradually erode their winnings. Indeed, such individuals just do not seem able to inhibit their response to the large payoff so as to capitalise in the long run. They seem to have completely lost their sense of conscientiousness.

We now have a basic understanding of just how some of the basic tendencies are modulated in the brain. However, what determines one's unique levels of reactivity or inhibition? The findings of recent times suggest that the five factors are highly influenced by one's genes. This might sound counterintuitive in light of the claims of the previous chapter, which argued that the effects of genes and environment couldn't be separated. Nevertheless, while it is not possible to draw distinctions between the influence of genes and the environment upon the development of a single individual, it is possible to infer the influence of genes on a specified characteristic across a wider population. The technical term for such studies concerning degrees of influence is known as the study of *heritability*.

The typical method for extracting such data involves studying large sets of identical twins – who share identical genes, and non-identical twins – who share the same number of genes as do any other sibling. The differences between the two twin types are argued to represent the influence of one's genes. Using this method, scientists are able to estimate the degree of heritability for a given characteristic (although, this is quite different from suggesting that any particular gene is the origin of a trait). In this way, it has been possible to show that the amount of variation across the sampled population for each of the five factors lies somewhere around 50 per cent.[11] Thus, elite athletes certainly inherit at least a portion of their mental fortitude.

There is little to be done about one's genetic heritage in hindsight. Accordingly, it is perhaps preferable to move onwards, particularly when one notes that around 50 per cent of the variation in personality must still be mediated by one's nurture. With this thought in mind, it is time to explore some of the environmental influences on a sporting personality that have been implicated in recent years/decades. And where better to start than at the beginning of life.

Echoes from the womb

The life of an infant begins not in the wider world, but in the mother's womb, where an enormous amount of crucial development occurs. Nevertheless, until relatively recently remarkably little research had been done concerning the life of the unborn child. This was at least in part because of the lack of available technology with which to observe and record the environment of the womb. However, a prevailing attitude alluding to the notion that the most critical developmental stages occur in postnatal life may have contributed as well. In any case, recent decades have seen remarkable advances with regard to the technological limitations of previous generations. The invention and refinement of techniques such as MRI scans have revolutionised the way in which we look at the earliest moments of life, in turn deeply affecting the way we view prenatal developmental. It has become clear that the unborn infant reacts to his environment, albeit in the limited sensory and cognitive capacity available to the unborn child.

In 1986 a well-reported and influential study discovered that learning and memory was possible even before birth.[12] Anthony DeCasper and colleagues asked expectant mothers to read aloud the well-known Dr Seuss story-rhyme 'The Cat in the Hat' twice a day for six weeks before the scheduled birth. Shortly after the birth the babies were tested with pressure sensitive dummies that triggered recordings of 'The Cat in the Hat' rhyme. The infants who had been exposed to the rhyme in the womb sucked considerably more vigorously, suggesting they remembered the song from the womb. The song was played both with and without the use of the mother's voice, further implying that the song, and its particular rhythm, had been learnt and not just the mother's voice. More emphatic has been the finding that unborn infants react (evident by the observation of increased heart rates) to rhymes that their mothers have recited repeatedly over a period of time, illustrating that even the unborn child has a degree of learning and experience in this early developmental stage.

What research such as this reveals is that the prenatal infant is not immune to experience and learning, merely waiting for birth to happen in order to begin the developmental genesis. Critical events in the brain of the unborn child are unfolding and will be likely to manifest themselves in some regard later down the line. This will of course have an effect upon the developmental trajectory for all children, including those who later become elite athletes.

With this research in mind, there have been a number of studies concerning the psychological implications of the prenatal conditions for the unborn child. In studies of rats it has been seen that maternal stress is correlated with an increased level of anxiety for the offspring.[13] Typically, these rats are less explorative of their local environment, which suggests nervousness of new experiences, and show increased defecation, another sign of anxiety and neuroticism.

This has also been found in human infants.[14] Where mothers have reported stressful pregnancies, links have been shown to suggest that subsequent temperamental difficulties might exist into the third year. Some have even linked these early difficulties to later adult psychological conditions,[15] although it is fair to note that the effect size of this influence was small.

While this early, and oft-ignored, aspect of development is surely but one of a large number of pieces in the wider puzzle, one should not be too hasty to discount the influences held within the womb for a sporting success story to emerge. Indeed, this seemingly innocuous feature of the environment can be seen to foster (or not as the case may be) the early stages of an elite athlete's development. As we have seen, openness to new experiences and low neuroticism are key assets in an elite athlete's armoury. If an infant's early behaviour is characterised by extreme cautiousness and neuroticism, there may be throwbacks for a significant period afterwards. With this in mind, let us move towards the early months and years of life for further clues to the origins of the personality of champions.

Talking without words

Until the late 1970s the consensus view among developmental psychologists – and still prevalent among lay folk – was that the infant child was just a tiny, helpless entity, with little control over the world around it. This is an entirely understandable position to adopt. After all, an infant is completely vulnerable to the environment without the watchful eye of his parents. And so it would seem logical that when one considers communication between parent and infant, the verbal traffic would presumably happen in one direction; from adult to infant. Yet this folklore was given a radical overturn following a series of seminal studies.

One of the earliest of these infant studies was conducted by paediatrician Berry Brazelton and colleagues.[16] Each infant in this study was seated with the mother positioned in front of the child at close proximity. The facial expressions of both mother and infant were ingeniously committed to tape by positioning mirrors at the appropriate angles, meaning that the expressions of both could be viewed simultaneously in the recorded footage. This allowed the researchers to observe the subtle nuances of the interaction with both mother and child visible on the same television screen.

The subsequent footage proved to be highly intriguing. Rather than the mother autonomously leading the communication, a far more complex relationship between the two was apparent. If the mother was attentive and playful – cooing at her child – the infant would often respond symbiotically. However, the infant was also prone to turning away from this attention and focusing on something else altogether. So far, this might sound distinctly as if the infants were simply without the capacity to engage in a structured mode of communication. But what became clear in the video analyses was that the infants appeared to actively modulate the level of attention from their mothers by their selective interest in her attempts to communicate.

This was elegantly demonstrated by the observation that when a mother was asked to be placid and emotionless in this face-to-face scenario, and thus preventing the infant from modulating the flow of the interaction, signs of distress such as frowns and grimaces were observed from the infant. And so, it was shown that the infants were not mere passive viewers drawn to the most salient attraction in the world around them.

In light of such studies these early patterns of relating were explored in order to identify if problematic relationships between mother and child could be pinpointed to a non-harmonious form of early communication. The findings suggested that this issue was at least part of the problem for a set of families with clinical problems concerning their infant.[17]

It is common to hear of such troubling mother–infant interactions and these difficulties have been suggested to lead to unfavourable developmental trajectories for the child. Indeed, if a child is perceived as being 'difficult' or 'stubborn' – by virtue of the communication breakdown between mother and child – the mother (as well as others) may respond by treating the child in a manner that 'befits' such a disposition. This of course sets off a chain of events that would certainly not be desirable.

There is the additional issue of what the implications might be if an infant is denied the opportunity to develop their sense of autonomy through the medium of 'manipulating' the adult. Indeed, the issue of self-reliance hitting an early developmental stumbling block – a highly desirable quality for all elite athletes – has already been suggested to be a possible implication of such relations. The upshot of such research is that while a problematic parent–infant communication might go (formally) unnoticed, the impact upon future development might be profound.

With these more covert influences in mind, an exploration of those more recognisable overt life influences is called for and the role of the teacher or coach is a good place to begin.

The Pygmalion effect

The message from the accounts above clearly emphasises the very different environments in which developing infants might find themselves, even from their earliest days. But surely the school and sports training environment are different, with the very role of a teacher or coach existing to maximise all students' potential, right?

In the late 1960s, the psychologists Robert Rosenthal and Lenore Jacobson became interested in exploring whether pupils whose teachers possessed higher expectations performed better than those of whom less was expected.[18] They devised an experiment to test whether or not implicit communication was being transmitted to pupils on the basis of their perceived intelligence, and in turn whether this would have an impact on their academic progress. In the initial stages of the experiment, all pupils in an American school from grades one to six were given what was termed the Harvard Test of Inflected Acquisition, near the start of the academic year. The teachers of these pupils were told that the test was a measure of an individual's likelihood of having an academic spurt within the coming months. They were also informed of those pupils who were deemed to be the most likely to make these significant academic improvements over the course of the year.

The outcome of this experiment was that the pupils who were said to have the greatest potential showed higher scores on IQ tests administered a few months later. However, the reason behind the improvement was a fascinating one. Indeed, rather than testing the children for potential to produce an academic spurt, the experimenters had given the children an IQ test and nothing more. Furthermore, the children were randomly assigned to the 'likely to spurt' groups. As such the pupils the teachers believed were more likely to make significant progress were actually just a random cross-section of the children in the study. In other words, what was discovered was that the perceived ability of the pupil in the teacher's mind was an important factor in the progress of the child.

The impact of such research is of course considerable, both within the schooling environment, as well as in a sports developmental setting. Rosenthal and Jacobson were witnesses to the occurrence of a self-fulfilling prophecy, wherein the 'able' pupils were made more so, whereas the 'less able' pupils were, at least in an indirect and presumably non-malicious manner, neglected. This is a worrying finding because the aim of any school or sports programme is to nurture and improve a student or athlete to their maximum capabilities, irrespective of their starting abilities. What is of greater concern within an elite sports setting is the notion that a coach may direct these forms of implicit communications towards a player who is perceived (perhaps erroneously, draw your mind back to the opening chapter) to be talented for aesthetic reasons.

Unsurprisingly, this important finding has been adapted by sport psychologists who have outlined a series of steps that underpin the expectation-performance process.[19] Firstly, the coach forms an impression of the player based on factors such as body size, past performances, perceived talent, and gender, among many others. Clearly this impression will be subjective and unique to the coach in question; people often see the same thing in strikingly different ways. These impressions and/or expectations of the child's sporting abilities are then pivotal in leading the coach towards a particular style of communication with the young athlete. A young boxer who is (perceived to be) less physically gifted might be subtly encouraged not to worry if they can't manage a particular technique or the final few press-ups in a training session – *'don't worry, by next week you will get it'*. However, those perceived as being more able would be pressed slightly more extensively in order to produce the answer that was expected to be within their capabilities – '*come on, you know you can do it. You're not stopping now*'.

Accordingly, in forming these impressions and acting upon them – often unwittingly – the coach begins the trend of differential learning within the training group. And it is this differential treatment that helps to form the young athlete's sense of self on the sports field, drive and motivation, and so on. Perhaps unsurprisingly, as these concepts are subtly reinforced upon the young athlete, a self-fulfilling prophecy begins to emerge. The youngster begins to embody the characteristics held of him by the coach and so the cycle continues.

So what, some might say. After all, don't we want the best young talents to be treated differently so that their abilities can flourish? Perhaps so, but the first chapter should have made clear that the initial impressions regarding abilities are not always accurate. Accordingly, the danger of such coach expectations is that an aesthetically unpleasing athlete might suffer the subtle sentiments that he is not quite good enough; that others are just that little bit better. Again, this need not be an overt manifestation of any biases held by a coach. They may arise through near-imperceptible modes of communication. But as Rosenthal and Jacobson have demonstrated only too clearly, these subtle influences can really make a pivotal difference in how the athlete perceives himself and of all that arises off the back of such self-perceptions.

The effect of such interactions can be seen in the early development of none other than NBA king Michael Jordan when he failed to make his high school's starting line up. However, Jordan was fortunate in having a coach who, rather than neglect him as just another also-ran contender, picked him up every day and drove him to school, took him to the sports hall at lunch, and repeatedly reminded him '*you have to stay focused, MJ*'[20] so that he might make the cut in future seasons. Jordan seems to have been fortunate to have had a coach who suspended any reservation he had about his early ability and kept his expectations high. And that view was certainly prophetic, for Jordan became, to many, the greatest basketball player the sport has ever seen.

The natural query at this juncture is to wonder whether these nuances exist in other settings outside the remit of the coach. Indeed, what role might the dynamics of the family hold for shaping the parameters of the sporting mind?

Birth order and you

For many millennia the family has been seen as a crucial component of the emergence of an individual's character. But in holding this view many people have presumed that the family is a generic entity that imparts the same form of influence upon each of its young. Research over the last century has illuminated just how far from the truth this assumption might have been.

Consider the following. Firstborns typically possess higher IQ levels[21] (although this claim, like many concerning birth-order effects, has been controversial). The firstborn child is on average a higher wage earner and more highly educated than his or her later-born siblings.[22] Most prominently, a high proportion of Nobel-prize winners have been firstborns.[23] However, for all those later-borns out there, frustrated at this apparent predicament, consider Frank Sulloway's[24] claim that the majority of revolutionary thinkers and activists throughout history, and accordingly many of the world's great folk heroes, were laterborn. Sulloway has trawled through hundreds of biographies to come to the conclusion that later-borns are more likely to challenge the status quo of society, whereas firstborns show a greater tendency towards orthodox values.

And this divide has also been applied to personality. Sulloway has argued that the five factors of personality – openness, conscientiousness, extraversion, agreeableness, and neuroticism – show significant birth-order effects. Firstborns are said to be less open to new experiences and more anxiety ridden. However, they are typically more extraverted and assertive, as well as possessing a greater degree of conscientiousness, being more responsible and achievement oriented. Conversely, laterborns are less conforming and unconventional, as well as more adventurous (relative to their eldest sibling).[25]

These contrasts in outcome between siblings of the same family are striking because they suggest that the family is far from the homogenous environment where all share the same experiences and upbringing. One would not expect distinct difference to emerge across a wide sample pool if the conditions faced by each child growing up were broadly equivalent. Naturally, the question that begs to be asked is why this curious phenomenon of firstborns achieving greater success in areas such as academia, business, and politics – with later-born siblings appearing to thrive in more radical or creative settings – should occur? The finding cannot lie in the realm of the gene for each sibling's genes are randomly allocated from the parental line; there would be no way in which to detect the firstborn from a later-born by means of a genetic test. The issue of an accumulation of creative genes in later-borns, and likewise genes for responsibility and acumen in firstborns, is simply a misnomer. The origin must lie somewhere else.

Research into birth order positions occupied by various siblings of any given family has illuminated some potential reasons behind such findings. One example of these differences is the nature and extent of attention bestowed by parents upon the children of the family depending on their order in the birth rank.[26] The first child holds a special place in a large number of societies, with lines of royalty passing through the eldest heir; the eldest sibling often takes on the family affairs in the event of a death, and so on. Accordingly, a common parental objective for the firstborn, largely modulated by societal desires, is to create the perfect child who achieves highly academically, as well as in other areas of life, sport notwithstanding. The child is bestowed with all the plans and dreams of the parents, and is often subjected to a regimented upbringing.

This need not be as sinister as it may appear at first reading, and indeed may well have many benefits for a young infant. It might simply mean that the child is brought to the doctor more often for modest complaints and encouraged to walk and talk from an earlier age than the subsequent later-borns. Accordingly, greater pressure to conform is often noted with firstborns who are placed into a more disciplined arena when compared to their later-born siblings. This may explain much of the tendency of firstborns finding themselves in positions of responsibility; they have simply been subject to this climate for so long it feels most natural.

Another significant difference in sibling environments is evident in the type of contact experienced by siblings through the birth order. Whereas later births will tend to spend much of their time with their sibling/s, firstborns will for a portion of their life (that is, until they are joined by their later-born siblings) experience a very different social landscape.[27] Rather than a world occupied by other infants and children, the firstborn is thrust into an adult world filled with very different challenges and experiences. For example, firstborns, in receiving more adult attention and exposure, are effectively being 'hot-housed'; in other words, they are receiving a wholly natural dose of intensive mental stimulation simply because of the position of their birth order. The expectations for an infant to rise to the occasion in an adult-only world will of course be significantly higher than for one whose world is constructed with a mix of adults and other infants. The Matthew effect – described in the previous chapter – illustrated how the small differences that exist between individuals can lead to significantly different life courses over time.

These differences occurring from one's birth order will of course impact on an individual's sporting makeup. As we have seen above, certain mental characteristics appear time and again in the mental makeup of elite sportsmen – dedication, conscientiousness, low anxiety, and emotional stability, among others. The experiences that familial birth orders prompt have commonly shown links to a number of these characteristics. High levels of motivation, discipline and responsibility have all been associated with firstborns on account of the high degree of attention and expectation heaped upon them from an early age. These traits might help firstborns to train more rigorously and regularly, and therefore develop expert sporting skills. Conversely, in sports where great cohesion is required, as in any team sport, the later-born's tendency to be malleable and sociable is an asset, although leadership roles in teams may be more frequently occupied by the firstborn more comfortable with the weight of responsibility and expectation upon his or her shoulders.

Such a claim may invoke a degree of distaste from those whose birth order does not fit with their projected or idealised view of themselves. Are there no exceptions to the rule? What should be made clear is that this birth-order effect is not set in stone for a number of reasons. As an example, a later-born may simply excel over his elder sibling in sports through sheer good fortune of physical stature. Also, in many families a later-born might well receive *more* parental attention throughout growing-up, as they are always the 'cutest'. Several other factors may change circumstances as shaped by birth order. Divorce, family deaths, and adoptions are just an example of a number of such possibilities. Indeed, if a child who was the youngest should experience the death of his elder sibling he will find himself in the place of firstborn. In fact, the loss of one child may spur the parents to make even more of a success of the remaining child and thus bestowing greater expectation upon the child than might have been experienced by the original firstborn. The converse might be true if a stepbrother or sister is introduced to the family following a re-marriage which converts an elder sibling into a later-born. The crucial thing for one to be aware of is that the trends are not witnessed because of the mere sequence of one's birth order; instead it is the functional outcome of such positioning that may be crucial. Earl Woods, father of world number one golfer, Tiger, has stated that the news of his impending fatherhood following a break of many years instilled a tremendous sense of wonder in him, *'why me, Lord? What have I done to deserve this beautiful and special kid?'*.[28] And so, in a functional sense Tiger Woods could be argued to have been fortunate to receive the attention that often is reserved for the firstborn.

The notion that birth order might play a part in defining the developmental trajectory of an individual is an intriguing claim. However, birth-order effects have not been without their detractors, even with the qualifications outlined above. As one shall see shortly, a considerable exchange of words has been voiced in recent years on this matter of birth-order effects and personality.

Peer pressure

We have seen the framework for the argument that birth order, and implicit family dynamics/politics, plays a part in forming an individual's personality. However, not all agree. The notion that birth order plays a major role in the formation of personality has come under critical review from a number of sources in recent years, most notably from the independent scholar Judith Rich Harris. Harris has reviewed a large collection of research findings in the realm of behavioural genetics (such as the twin studies of heritability mentioned earlier). She has claimed that rather than the family socialising the child and defining their personality, instead it is the child's peer group that is the more pivotal influence.[29] This contention has not been short of its critics, largely because, as Harris suggests, we have become so used to the family being held as the central piece of a child's development it is almost unimaginable that it does not hold the importance traditionally afforded.

Some historical background is appropriate at this point for the debate has raged for some time in its various guises. Early-twentieth-century psychoanalytic concepts (stemming from Freud and his intellectual descendents) placed the parent, and particularly the mother, at the heart of the developmental process. In fact, somewhat shockingly, conditions such as autism, now known to possess a strong genetic component, were often attributed to an un-nurturing, and/or aloof parenting style; the so-called refrigerator mothers.[30] Accordingly, parents of autistic children were often vilified, quite unjustly as we now know.

The advent of genetic science soon released such parents from the stigma society deemed to place upon them. But this often meant that the blame was attributed at the other end of the spectrum; now not with nurture, but with nature. Genes were viewed as the be all and end all of human outcomes (the jury is still out on conditions such as autism). However, throughout these nature–nurture battles not too many stopped to think that maybe a different form of environmental experience besides that of the parenting might be of greater influence. The notion that the child's peer group might just be the most pivotal of all influences upon personality was certainly not a serious consideration, at least until recently.

Upon reflection this idea has much to say for the matter of psychological development. After all, we evolve to be successful in the environment we find ourselves in – this tenet being the fundamental basis of Darwin's theory of natural selection. And it would seem more sensible for the kinds of traits one develops to be gleaned from the arena where success in later life is most crucial. Accordingly, it would seem to be likely that a child's peer group shape a portion of their personality, so that they might be able to form behavioural tendencies conducive to their survival prospects in the real world. An example of such peer influences can be seen when immigrant children quickly assume the language and accent of their peers, instead of keeping the accent of their parents.[31]

The logic of the Harris hypothesis has been supported by a large body of work in the realm of behavioural genetics suggesting that, contrary to popular belief, the home life and parental styles are not the major differentiating factors behind the personality of the child. (In fact, Harris makes a qualification of this point insomuch as she argues that the personality of the child within the family may be influenced by family dynamics such as birth orders, but that the personality elsewhere is not influenced by these factors). The evidence that stands in favour of such a notion has been held up to scrutiny over the last decade or so and shown itself to be both robust and reliable. For example, when identical twins have been measured for alikeness, they are no more alike whether they shared a home or were raised apart.[32] The implication of such a finding is that it does not much matter which home one finds themselves growing up in; personality is not strongly defined by one's home life. One's genes will still of course exert their influence. As we have already seen, they can explain approximately 50 per cent of the variance in personality between individuals. The really curious finding, however, is that the remaining 50 per cent seem not to be accounted for by one's home environment. The shared environment of the family does not seem to add to this inherited portion of personality.

One rebuttal to this finding has been that perhaps the effect is forged through the false assumption that the home life is the same for all offspring. This, after all, was the very argument addressed in the previous section. Perhaps rather than a single homogenous environment, instead a series of microenvironments exist that sculpt the child accordingly? This is in line with the birth-order inferences but as you may be thinking, something has to give at this point. And it would seem that the birth-order prophecy is the weak link.

This assertion is possible because a carefully conducted study taking into account these possibly confounding microenvironments of different parental treatments towards their children was implemented to answer just such a question. The study made sure to take detailed measurements of differences in how individuals were treated within the family setting – such as a greater weight of expectation on firstborns, and so on. Accordingly, this information could be used to check if this form of influence was effecting the results. It was not.[33]

This is a profound finding as it strikes at the very heart of much of the child-rearing advice parents are frequently exposed to, perhaps rendering much of it a waste of time and money. More controversial has been the counterclaim that if this is the case then why does child abuse in the home cause such damage down the line, and why even bother about one's offspring if it really makes no difference anyway?

These kinds of questions are almost certainly an overreaction to new and counterintuitive findings, as has often been the case in science. Parents and home life are absolutely important in the healthy upbringing of any child. The claim is merely that the non-genetic personality differences that are found in individuals are more likely to have emerged outside the home where, as we have seen, the individual will need to stand on their own two feet. Thus, the evolutionary benefit of such a method for sculpting personality is that the child is able to find an appropriate niche within the wider social world, and not just within the family. This is not to deny the very fundamental needs that the family environment is vital in providing; love, emotional warmth, and so on. It is probable that the family does contribute in some way even if this is only to prompt the kind of peer groups a child is drawn to, but this would perhaps be a moderate assessment.

For the young athlete, the argument of Harris has interesting implications. The suggestion that one's peers are a more important influence upon one's personality means that the right peer group is a crucial component of the elite athlete's development. This is not so surprising really. After all, it is normal to group children based on ability. And while this is at least in part so that advanced skills can be taught to those gifted youngsters, it is likely that the influences of one's peers, who share similar abilities and aspirations, will have its own effect.

The proof of such an assertion might be argued to lie in the observation that elite athletes often emerge from environments where they were not the only one's striving towards the goal of elite status. This can be attributed to the high quality training such athletes might have received over those who were less fortunately brought up in a less conducive environment. However, the influence of a peer group, each striving toward a common purpose – the attainment of elite sport status, will certainly play a pivotal role. This seems to have been the case for the young Mike Tyson who was rescued from a life in and out of New York's young offenders institutions. The famed trainer

Cus D'Amato recognised Tyson's potential and brought him to live in his mansion home, which essentially doubled as a boxing academy with many other young prospects in residence. And the same can be said for the tennis academies that have produced so many of the world's top tennis players; Andre Agassi, Monica Seles, Jim Courier, Andy Murray, and Maria Sharapova, among so many others.

Success builds success

Thus far we have seen a number of the influences upon the developing character and of how they might relate to future elite sporting success. However, one of the most basic forms of influence upon a young athlete's motivation, confidence, and drive – his experiences of competition – has not yet been discussed. Indeed, as many will be aware, there is little to drive the enthusiasm of a young athlete like the feeling of a podium finish in a tournament.

The sentiment that success builds success is one of the corner stones of Albert Bandura's self-efficacy theory,[34] a framework for understanding how an athlete arrives at the sense of confidence in his abilities (although it should be said that Bandura's work was not specifically designed with sport in mind). Indeed, performance accomplishments have been shown to predict the degree of perceived ability of an athlete. This is not wholly surprising. After all, those who win often are likely to realise that they are pretty good, while those who lose often perhaps will recognise that others seem to be a whole lot better than they. Indeed, tennis superstar Pete Sampras has illustrated in words what a lot of people are intuitively aware of; *'winning any victory gives me a lot of confidence'*.[35]

This most basic of observations is in many ways more profound than it is often given credit for. Most are well aware of the important influence that success has on the self-esteem of a developing young athlete (not to mention the rest of us). However, it is remarkable how often coaches and parents in their overeagerness to expose their charges to competition the grade or age group above their own overlook this basic principle. Indeed, this state of affairs can often become a syndrome in youth sport where the prestige of playing 'up' becomes as important as competing well in one's natural division. This is often for very valid reasons – a promising young player might well need to face stiffer competition than their own age group can deliver. Yet the possibility exists that in the rush to develop an athlete's game, self-confidence is dealt a blow as one is defeated by older and stronger opponents.

While experiencing regular success is an undoubted boost to the confidence of any young sportsman, too much of a good thing can also be detrimental. Indeed, while victories help to create a sense of efficacy concerning one's abilities, the lessons learnt through losses, however painful, are among the most poignant in sport. In the words of the legendary American football coach Vince Lombardi, *'if you can't accept losing, you can't win'.*[36] What Lombardi alludes to in this quote is that the recognition of one's vulnerabilities – faulty techniques, ill-advised tactics, poor fitness, or a wandering concentration – can be made and improved upon, often in a way that simply does not occur after a win. After all, who really criticises themselves too deeply after a win?

The occasional loss can also provide the young athlete with a reality check from time to time and remind them that losing is normal and inevitable even for the very best. After all, who can easily name an athlete who truly remained unbeaten throughout a career? For those who achieve early elite success – a national team place, terms with a major football club, and so on – the pressure from all those around to succeed can be overwhelming.

For one exceptionally promising young tennis player it seems that this extreme success was an unshakable burden. Al Parker has been recognised as the greatest junior player in American tennis history, with 25 junior national title victories, and many victories over future grand slam champions and top-ten players. Yet Parker did not even make it into the top-250 in the men's professional game, to the surprise of many. Part of the reason may have been Parker's prodigious early success. As a friend of the family has commented, '*I honestly think that one of the worst things that can happen to a kid in junior tennis is to be unbeatable in the 12s because he's always the one the pressure's gonna be on*'.[37] Parker himself has noted that '*potentially some of my success early one, I think, was a curse*'.[38]

With these observations in mind, any young athlete's development will require the careful balancing act of healthy doses of success, alongside the necessary losses that foster an understanding of one's competitive limitations, as well to allow one to be 'normal' and lose once in a while.

Conclusions

We have seen in the sections above just how complex the development of a personality is, and it would not be an overstatement to say that we have barely scraped the bottom of the barrel in this chapter. For a young athlete, the road to an elite sporting personality may begin with favourable basic tendencies, high extraversion and conscientiousness, and low neuroticism. From these beginnings the shaping process begins in earnest. Our early experiences in the womb and interacting with caregivers can trigger detrimental behaviours, such as anxiety. Our peers, teachers, and parents influence us in numerous ways. We profit, or suffer, from our experiences, as Bandura's self-efficacy theory has shown. And all these factors are thrown together to create a rich mixing-pot of development that is wholly unique to the individual athlete.

As we close this chapter, let us think ahead to the next key component of an elite sporting mind – one's perceptual and decision-making skills. The following pages will take us on a journey through the 'eyes' of an elite athlete; how they see the playing field and make the best play decision, often in mere milliseconds. Is there a magic 'eye' for the ball, as some would have us believe, or is there more to the making than at first sight? Chapter 4, 'What You See Is What You Get', will answer these questions and more.

4

what you see is what you get

Recall, if you will, the description in the opening stanzas of chapter 1 regarding the returning skills displayed by Roger Federer while under the stern gaze of the Andy Roddick service delivery. Such prowess is of course a fine illustration of marvellous reactive abilities and split-second decision-making. Federer seems to simply have more time than his nearest rivals to prepare for and initiate his strokes. In turn, his nearest rivals are similar racket-wielding magicians of time when compared to the mere mortals of tennis. And the trend is observable among elite athletes from any number of sporting disciplines. From basketball and football to rugby, boxing, and badminton, one can witness crucial decisions made in mere fractions of a second occurring time and again throughout the course of a contest. Trevor Brooking, director of development at the Football Association, emphasised this crucial point when he asked of the young English footballing talent, *'Have* [*they*] *got the know-how to make decisions? Do* [*they*] *know how to use it and select the right pass?'*[1]

We routinely marvel at the finest competitors for their ability to sense the opening in the opposition's defence, to slip and slide away from the stiff jab, or to spot the availability of their teammates for a pass and to make the appropriate response. With regard to the power serving of Andy Roddick, his opponents have less than half a second to organise their response before the ball is upon them. To put this in context, from the start of the sentence to this very moment, Roddick would have delivered his service and Federer would have already returned it. No mean feat in facing and disarming this colossus then. So, how do great athletes realise these great abilities? What allows such sharp reflexes to emerge time and again allowing the perfect shot for the situation to emerge under such time constraints? Above perhaps all other mental abilities these kinds of feats are typically considered innate gifts – *you've either got it or you don't*. By now one might be more sceptical of such a black and white view. Indeed, there is more to these feats as one shall see. To answer such questions let us begin by exploring a brief portion of what is known of the visual processing system.

The magic eye

Of all the sensory systems, the visual system has been by far and away the most studied. This is not surprising when considering just how much of the human world is based on visual abilities. We routinely read books and magazines, drive cars, watch television, and, as I am doing right at this moment – type on my personal computer in order to meet the deadline of my publisher. These are activities all wholly unavailable to one without the faculty of sight. Although we mostly take this ability for granted (it is a ubiquitous feature of daily life after all) it only requires a brief moment of thought to appreciate just how remarkable this gift of sight really is. In fact, this ability is so remarkable that its very existence has prompted a longstanding debate between religion and science, the ilk of which exists to this day.

William Paley, an early-nineteenth-century theologian, studied what was known of the eye in his day and came to the conclusion that anything so complex and perfectly adjusted to its purpose simply must possess a creator. He likened this design process to that of a fine watch, where the precise details of the dials were sculpted and realised by a master craftsman, and claimed that this was evidence of the existence of a God.[2] And in spite of Darwin's theory of Natural Selection, which has since shown with some elegance just how an entity such as the eye could have evolved without the design plans of a creator, this appeal to divine origins prevails in certain quarters.[3] In any case, the somewhat 'magical' quality of sight remains and the visual system should certainly not be seen with any less wonder. And the eye is also of undoubted importance in the matter of elite perceptual skill for sport champions. So, how does vision work?

Democritus, a pre-Socratic Greek philosopher born around 460 BC, was one of the earliest-known investigators of the phenomenon of sight who attempted to provide an answer to this question. He believed that the objects we habitually see around us, tables, chairs, postboxes, and so on, possess a kind of 'aura' which he termed the *idol*. This idol was supposedly transposed whole and unbroken, rather like a ghostly template or shadow, from the object to the eye, where it would be received and viewed.[4] Unsurprisingly, modern accounts heavily contradict the account of Democritus, but let us not judge him too harshly. While the advent of modern physics has been available to us to exploit for some decades now, Democritus was, of course, forced to rely on the rather cruder methods of analysis reflecting his earlier times.

Such earlier accounts aside, visual science has now forged a considerably more empirically tenable account in recent decades. The very beginning of any sensory processing occurs when stimuli reach the sensory receptors of the body. In the case of the visual system, these receptors are, of course, one's eyes. The stimuli that eyes have evolved to work with are waveforms of light. These waveforms are received, transduced, and finally coded into a medium that the brain can manipulate. Put simply, the near-infinite arrays of light patterns that bombard one's retina are converted into electrical impulses that result in corresponding brain activity. These electrical impulses are sent to a part of the brain called the visual cortex where the many qualities of the visual world such as colour, shape, and motion, are processed, albeit in different locations for the various aspects of vision.

With this basic information regarding the functioning of the visual system one might expect the elite perceivers of sport to possess a host of visual advantages in line with their great abilities in handling rapid plays from the opposition. Many possibilities, some more incredulous than others, have been mooted in the past. Perhaps elite athletes have eyesight better than 20/20 vision. Maybe more photoreceptive cells exist in the retina of the champions allowing them to render a finer representation of their opponents' deliveries thus aiding them in their judgement of the flight path? The baseball legend Babe Ruth was claimed to owe his record-breaking homerun prowess to the fact that his eyesight was reputedly beyond that of the normal individual[5] and so he was able to see the ball approaching his bat with greater clarity and precision.

Despite such claims affirming these popular notions, empirical tests have failed to find a single case of elite athletes possessing greater visual acuity than the novices of sport. Many of the historical studies that have filtered through to the popular press over the years have been shown to have arisen either as a result of poor experimental procedure or as simple hoaxes staged for the benefit of an (often unscrupulous) individual's wallet by claiming some form of visual acuity enhancing hocus-pocus. It was seen in a study conducted in the United States that 15 per cent of players in the National Football League (NFL) and 20 per cent of players in the National Basketball Association (NBA) had *poorer* than average visual acuity.[6] Similar findings are observed when one explores dynamic visual acuity, that is to say the ability for tracking the pathway of a fast moving object. And so it can be claimed with some confidence that elite athletes, as a general rule, are not visually advantaged when compared to the novices of the sporting world.[7]

This point is further illustrated with some elegance by two examples, one anecdotal, the other scientific. The first example involves the legendary basketballer Michael Jordan, who was said to have made free throw shots in practice with his eyes closed, before turning to his junior teammates and stating provocatively, '*welcome to the NBA!*'[8] Jordan is demonstrating, besides the status his skills have bestowed upon him, that great visual advantages are mere appendages to the wider bag of tricks. After all, he apparently didn't need the full benefits of vision to pull off at least a portion of his skill set.

However, one might argue that this simply represents a well-rehearsed movement that no longer requires overt visual attention to be successful. The dynamics of open-play are surely a different scenario entirely? And so, it is with some interest that we move to the second example, for it concerns the discovery that cricketers (and the same is thought to hold for all fast ball sports) do not actually look at the ball while it is travelling through the air towards them.[9]

This seems remarkably counterintuitive at first, particularly when one considers how, invariably, the first lesson that any young player will have learnt playing tennis, football, golf, or cricket would have been to watch the ball. Instead, what the experimental findings have shown, with the aid of head mounted eye-tracking devices, is that while batsmen appear to focus upon the details around the release of the ball, they then rapidly move their focus to a point near where the ball is likely to bounce.[10] For the main part, the actual watching of the ball seems to be a distinctly underutilised process in the whole batting procedure.

Echoing the implicit sentiments of Michael Jordan, pure visual qualities are certainly not the be all and end all, a fact that does not seem to have dented the multimillion dollar business concerned with enhancing sports-vision. This is not to say that good, functional vision is not of benefit to athletes at all levels of sport. Any sportsman with severe short- or long-sightedness will obviously be less able to make sharp visual distinctions at ranges outside their visual comfort zone. A tennis player with uncorrected short-sightedness might struggle to detect the finer stroke play of an inventive opponent. They might also avoid approaching the net as the baseline may well offer a period of respite with which to track the ball more clearly. Nevertheless, the fundamental observation of visual potency in sport is that there have been no mean observable differences between elites and novices; rather, what seems to be evident is that there are those with great and poor visual acuity among both groups. The true differences in ability must lie elsewhere.

The magic number seven

Clearly the advantages in reaction times are not garnered by the scope of the elite athlete's vision and eyesight. Thus the question that begs to be asked is how the best perceivers use the visual information that they receive to greater effect? Indeed, if they do not fundamentally *see* any better they must surely be *using* what it is they do see considerably more efficiently. Curiously, the early research into this area which has proved most revealing was not a study in the arena of sport *per se*, but instead in the rather more cerebral confines of the chess world.

Before we make further headway in this matter, a brief step into the realm of cognitive psychology (the study of mental processing), and in particular the study of memory, is required. George Miller, later to become president of the American

Psychological Association, made the observation that limitations exist in the amount of information one is able to hold in mind at any given moment. Working memory, the name given to this finite memory store, was argued to hold just seven units of information, plus or minus two to account for inevitable individual differences.[11]

Broadly speaking, this is an intuitively correct proposition (although there is some debate about the exact capacity of the working memory storage capacity[12]). One can test this premise simply by attempting to hold in mind a number of more than seven (plus two) digits for a short period of time. You can try this task if you are so inclined using the numbers shown below. But remember, once you have viewed the digits do not look back again, and after a brief while write down what you remember and see how good your recall was. The number to remember is:

7 – 0 – 8 – 5 – 3 – 2 – 9 – 4 – 1 – 6

You may have found this extremely difficult and most likely an impossible task. This number, after all, contains ten digits, well over the seven (plus or minus two) claimed by Miller to be the maximum capacity. It is simply beyond the capabilities of most people. Yet, a remarkable observation of elite chess players in the late 60s and early 70s illustrated feats of memory that appeared to debunk this received wisdom entirely.

Players were shown a chess board littered with the pieces of a game in progress (one they hadn't seen the earlier moves to) for just a few seconds. Grandmasters were typically able to reproduce the positions of the pieces (typically around 24–26 pieces – a mid-game position), with up to 90 percent accuracy. However, the non-elite players, as one might expect, came nowhere near to such powers of recall, instead bound within the limits set down by Miller's observations of working memory.[13] They struggled to place even half the pieces in the correct location.

For many this is a clear sign of an innate difference between those who achieve the lauded Grandmaster status and those who simply harbour a keen enthusiasm for the game. The Grandmasters appear to break the very rules governing the human mind that others are apparently bound by. Surely this is proof of genius (if ever it were seen) and evidence for an advantage beyond the realm of mere practice and discipline, right?

The preceding chapters will perhaps have made one more sceptical of such statements. Indeed, more illuminating evidence soon came to light regarding this spectacular feat of memory recall. Accordingly, further probing showed that the elite players were indeed demonstrating powerful levels of recall of *real* chess situations, such as the mid-game snapshots they were exposed to in the original tests. However, their recall of a board with randomly placed pieces – positions that had not evolved through the dynamics of normal play – was no better than the non-elite players who were tested. Their working memory was no greater after all. But then how had they managed to perform such marvellous recalls in the authentic chess situations?

The conclusion that researchers arrived at was that the Grandmasters were exhibiting a cognitive strategy they termed *chunking*.[14] To chunk simply means to house multiple blocks of information together as one unit. Accordingly, rather than exhausting the capacity of working memory (remember, seven items – plus or minus two) one is able to conserve resources and thus hold more information in mind at one time. And this is something we do routinely without paying a second thought to its existence. For example, one cannot normally remember a full random phone number, complete with an area code. However, if the area code is the same as that of one's hometown (or highly familiar for another similar reason, year of birth, pin number, and so on), this code can be treated as one chunk of information. We can therefore remember the whole number more easily. The key point in this argument is that information stored in long-term memory is utilised as a strategy to reduce processing load on working memory, and thus increases the amount of information one can keep in mind at any given moment. So far, so good. But how does this help us understand the elite perceptual and decision-making skills of sporting champions?

The upshot of this method of representing pieces on the chess board means that rather than computing the on-line, real time permutations for each position, the Grandmaster is typically drawing from a database of equivalent situations that exist in their memory bank. In essence, they are cross-referencing against the many thousands upon thousands of previously played or dissected scenarios in their database of experience. However, the calculations that the novice chess player must perform in real time, due to a lack of any (sizeable) database of situations to draw upon, are vast and draining.

To the Grandmaster this means that rather than seeing a confusing mess of pieces swamped across the board, they are is able to observe order among the chaos; a chaos that novices can only blindly stumble their way through. Thus, the ability of elite pattern recognition is understandable through the interaction between short- and long-term memory.

If one transfers this ability observed in chess Grandmasters to the arena of sport, it is easy to see where it becomes useful. Let us take football as an example. If a typical Sunday league 'hacker' was transposed to the heart of a Premiership midfield he would be certain to feel that the pitch was awash with movement and action as the professional players around him jostle to create openings and scoring opportunities. However, he would be unlikely to detect the complex patterns and structures of play designed to achieve these goals that exist at this elite level. And certainly nowhere near the level in which top-flight England professionals, such as Steven Gerrard or Frank

Lampard, would be able to do. The elite professionals see possibilities of attack, potential dangers, and opportunities long before those of the terrace faithful. Of all the elite players, perhaps Canada's ice hockey legend Wayne Gretzky has displayed the greatest visual awareness ever witnessed in world sport. He was routinely said to have known the state of play a few seconds ahead of time, always anticipating the next move or available opening. With this sentiment in mind, Gretzky has claimed he sensed other players more than he actually saw them. *'I get a feeling about where a teammate is going to be… A lot of times, I can turn and pass without even looking.'*[15]

The record books speak for themselves. Gretzky is the highest scorer in NHL history, and relevantly to the issue of sporting vision, he also holds the highest number of assists by any player to ever play the game. Intriguingly, Scotland's motor racing legend Jackie Stewart has also alluded to this ability, presumably despite the absence of any formal scientific understanding behind his views; *People call what I have 'reaction time': it's purely the consumption and deciphering of information and acting positively on the basis of it.*[16] Presumably this reaction time was a portion of the remarkable abilities that took Stewart to his three World Drivers' Championships.

This claim of enhanced pattern recognition has been supported with a wealth of experimental evidence which illustrates how elite athletes are privy to significant advantages over their novice counterparts when it comes to recognising patterns of play. And of course this general principle applies across all sports where patterns can be detected.

Viewing the footballer's visual array as shown in Figure 4.1, it is clear that the player approaching the ball will have a host of options flashing through his mind's eye. The question is of course which of these options will lead to the best outcome for the team? If he tries to beat the defender will he be thwarted

Figure 4.1 The footballer's visual array.

before he makes his strike? Perhaps a better choice would be to pass the ball to his teammate on the right who might have a clearer run at goal, or at least greater scope to make a successful play? In making the best choice, the player with a database of play patterns will be most able to deduce where the dangers are lurking, and when his teammates are able to make their breaking runs away from the auspices of the marking defender. An inbuilt sense for the unfolding of a given scenario will be the result of such long-term memory faculties and bestow upon one an almost mathematical precision in sensing who is able to reach a particular point on the field at any given moment.

It's not what you see; it's where you see it

A further ability related to this pattern recognition faculty has also been found in elite sportsmen. For a tennis player, even one of elite professional standards, the service delivery of Andy

Roddick is a formidable, even wholly impregnable, proposition. Yet Roger Federer seems to have time to spare; the whole event often seems to be passing by in slow motion to those watching from the comfort of the front room. The discovery prompted by the chess studies led some researchers, in addition to the work involving pattern recognition, to investigate if differences also existed in the areas or zones upon which the great receivers of sport (batsmen, tennis returners, and so on) typically focused their attention. Are there clues that only elite athletes are tapping into?

The method most commonly used to study such deployments of visual attention is typically referred to as an occlusion study. Near-life-size video projections of a server or bowler are presented to an athlete. However, the crucial details of the seconds following the ball leaving the racket/hand are omitted; the footage is ended shortly before this moment. The athlete is then asked to make an estimation of the direction of the delivery and the type of ball delivered (such as the spin imparted).

One can use such studies and gradually manipulate the point of occlusion, that is to say, gradually taking more detail away from the viewer, to learn what physical cues athletes use to understand the delivery they face. Indeed, engaging such methods it became evident that the elite receivers still maintained a good understanding of the properties of the delivery well before it had been completed.[17] They were aware of the properties of a delivery without even seeing it leave the strings of the opponent's racket (or hand, in the case of the bowler). However, novices were significantly less competent at this task. They required more visual detail over a larger portion of the technique to make a correct decision as to the likely direction and spin the delivery would take. Clearly, elite receivers are not using the same visual information as their novice counterparts.

Whereas novices mostly focused on the actual release of the ball from the strings or hands, the elite athletes were seen to be more focused upon postural cues, such as the movement of the torso, hips, and shoulders, in order to generate early signs of where the ball was going to be directed, or how it was to be spun.

This is not so surprising when one considers the speed of the professional tennis serve (or the pace of the cricket bowler, baseball pitch, and so on); in the twenty-first century the ball now travels so fast that it is simply not humanly possible to watch the ball leave the strings (or hand) and make a decision/reaction based on this information alone. Therefore, the need to use anticipatory postural cues become vital at the elite level. It is worth stating that the top players will still watch the ball being struck in order to guide their response. However, the purpose of this will simply be to confirm a motor response already initiated, and not to extract the information needed to know where to move by itself.

Decisions, decisions!

So, elite athletes are using subtle, yet importantly different strategies, than lesser gifted sportsmen who aspire (unsuccessfully) to reach similar heights. But this, of course, is not the final word on the matter of the visual skills on display in the great arenas of world sport. As one will be only too aware, not only do elite athletes appear to anticipate and track the incoming challenges of their respective sport more efficiently than their amateur counterparts, they also consistently make the best possible decisions under the pressure of finite time. And this is often with mere milliseconds at their disposal. How is this possible?

We have already seen the increased anticipatory and pattern recognition skills of the top performers in sport. One immediate upshot of such favourable 'shortcuts' is that the elite athlete is able to make a greater devotion towards processing the required decision. Simply put, more time exists to evaluate the possibilities even if one is referring to mere split-second windows of opportunity.

However, while this facet is certainly of benefit, this only explains a portion of the differences between the decision-making processes of elites and novices. Indeed, more processing time could also be interpreted as more time with which make a bad choice. After all, how many have had too long to make a crucial decision and chosen unwisely? What allows the elite athlete to know that their choice is the correct one? The crucial difference lies in the fact that elite athletes are able to draw from a mental catalogue of tactical scripts, known as *action plan profiles.*[18] Action plan profiles are essentially a database of template answers to sporting 'questions' (such as '*what shall I do now that he kicked the ball over there*?'). These templates are formed through the many months and years spent experiencing a range of implications for the different choices made in similar situations of the past.

Perhaps the best way to understand such a concept is through an example. If one imagines a junior tennis player with a powerful forehand but little tactical experience, one obvious outcome of such a style of play will be that the young fellow will routinely hurt the opposition by virtue of the sheer raw power of his forehand stroke. Accordingly, the first tactical challenge to this junior will be to recognise the damage he has caused to his opponent's position. If he recognises the patterns in the unfolding moments, such as the fact that his opponent is on the stretch and off-balance, he will be in a position to take advantage and anticipate the likelihood of a short ball opportunity arriving on his side of the net.

So far this situation is indicative of the pattern recognition detailed previously. But, the youngster needs to not only recognise his opponent's vulnerabilities; he also has to know the correct decision with which to exploit this moment of weakness in his foe. This is where action plan profiles become crucial. If his experiences are extensive enough he will be aware that a fine choice in this situation might be to strike the ball down the line and approach the net, in order to further apply the pressure upon his struggling opponent.

However, this young prospect, unfortunately for him, is not yet that astute and allows the player on the other side of the net to play a looping, defensive ball and thus claw his way back from the jaws of defeat. In essence, this junior is playing with the basic techniques of the sport, but with a restricted set of action plan profiles, and thus is highly limited as a competitor. Unsurprisingly, the very best competitors have been seen to possess complex action plan profiles, rich in detail and covering a vast range of possibilities that might potentially occur. In essence, they have all the angles covered within competitive open-play. They know what to do and when to do it.

However, there is a problem with such a method of play. It can become predictable if repeated too frequently. After all, if one initiated the same template-strategy time and again in response to a particular question, the opposing player/s may begin to adapt and turn the tables. Indeed, if the young fellow with the dominating forehand wises up to his great strength and does decide to venture forward with abandon he will surely taste early success. But if he opts for this tactic every time, his opponent may become wise to this ploy and hoist the ball high over his head for a winning lob.

One of the greatest competitive skills in sport is to be able to successfully evaluate one's opponent while in the heat of battle and move from plan a to b, or even from plan b to c, whenever necessary. This shift is made possible by the use of *current event profiles*,[19] which contain tactical scripts and allow an athlete to selectively use his action plan profiles when and where most beneficial. One example of this need for well-developed current event profiles might be seen with the young tennis hopeful mentioned above. Let us now imagine he has learnt to attack his opponent down the line whenever he sees him out wide and off-balance. This tactic may well be successful against the majority of junior competitors. However, if he should come up against a particularly fast opponent who enjoys running balls down and lobbing the ball high into the air, this tactic might be counterproductive; he may be playing into his opponent's strength.

The requirement here is to incorporate the knowledge from his action plan profiles and adapt it to the specific requirements of his opponent. Accordingly, he might now avoid hitting the ball down-the-line from time to time, and instead elect to go back cross-court to break up the rhythm of his rival. The upshot of a well-developed set of current event profiles are clear. Athletes are better able to deal with the changing circumstances of a contest. Unsurprisingly, this is a skill usually only observed at the higher echelons of sport.

One such example of these action plan and current event profiles in full display at the highest levels of world sport was evident in the super-middleweight world-title fight between Welsh Dragon Joe Calzaghe and his hardened Danish foe Mikkel Kessler. Legendary trainer Emmanuel Steward commented that '*just at the time when everyone thought that the fight was moving for Kessler, Calzaghe made an adjustment and came back and took over the fight completely.*'[20] Calzaghe knew how to structure his attacks, but more crucially, he also knew how to adapt his gameplan to an opponent who was not susceptible to the initial strategy he adopted.

Training the magic eye

The overriding sentiment from the preceding paragraphs is that elite sporting vision and decision-making skills are not simply a freakish anomaly and merely the domain of the elite, chosen few. Rather, they are, at least in part, skills that are learned and refined over a lengthy period of time until they appear in that effortless form one will surely recognise from the performances of the great champions.

Recalling the virtuoso anticipatory feats of Wayne Gretzky, ice hockey's hall-of-famer, has said that his anticipatory qualities emerged from '*fear. Growing up, I was always the small guy. When I was 5 and playing against 11-year-olds, who were bigger, stronger, faster, I just had to figure out a way to play with them.*'[21] And so perhaps Gretzky was inadvertently placed in a situation where he had to make best possible use of his perceptual and decision-making faculties, and so developed these skills in a wholly implicit manner. The question that begs to be asked at this moment is whether one can learn these skills in a more contrived setting? In short, can coaches and trainers teach their players these anticipatory and decision-making skills through specialised coaching techniques designed to replicate the years of skill formation that more typically arises from lengthy experience?

Unsurprisingly, this is not a new question, at least at the higher echelons of world sport. The sporting world is now so competitive in near enough all of the many athletic disciplines that any advantage is highly sought after and the research into such advantages is rapidly accumulating. The possibility that athletes might be able to make more advantageous choices in the heat of battle is too good an opportunity to miss. Additionally, if players could be trained (at least in part) by a non-physical procedure, improvements could be yielded in those athletes who are injured and unable to partake in normal training sessions. The possibility also exists to train athletes without having to risk injury as a result of overtraining. So, what has modern sport science got to say on this matter?

Two basic methods have been typical with such training endeavours; video simulation learning, and field-based learning.[22] Video simulation utilises video footage to expose players to specific situations that have pivotal status within their sport – such as the postural cues of a server or the patterns of punches a boxer typically delivers. Field-based learning operates upon broadly the same ethos of exposing players to key tactical situations, but takes place on the playing field of the sport in question. However, can either technique deliver the 'goods' and turn an also-ran into a player of substance, or are they simple razzmatazz?

The use of video footage has been the more common of the two approaches for developing sports perception and decision-making skills. A typical experimental set-up to test for postural cue-learning possibilities often involves a participant viewing a video screen (frequently at near-life-size) with a server/penalty-kicker/bowler facing the participant and executing the relevant technique as if the participant was truly facing him.

So far, this is pretty uninteresting stuff. After all, we see this kind of footage on the television all the time. However, as we have seen above with the method of temporal occlusion, the opportunity exists to expose the athlete-in-training to key patterns of play, or important postural cues during a technique, which might be used to predict the intentions of an opponent. And it has been shown that by using such methods improvements can be made in the predictions of the delivery's characteristics.[23]

This is a fascinating observation for it suggests that a whole new way of coaching athletes might become a very real prospect in the near future. Instead of goalkeepers simply facing penalty-kick after penalty-kick in order to train their defensive faculties, much of this training might soon be done in the laboratory. The nature of the kicker's posture could be broken down and the resulting information used as a performance aid. In other sports, one might well see the same happening as well. Instead of boxers spending hours on the target pads honing their reflexes and sharpness, they might elect to use some of this training time developing their postural cue awareness so as to enhance the very ethos of boxing; namely, to hit without being hit.

Intriguingly, in professional sport this might well start what is know as an 'arms race', wherein the advancement of one individual's abilities sparks off a chain reaction in others who try to match and exceed the development. Perhaps the penalty kickers would start to disguise the wind-up of their kick, or something to that effect. This is pure hyperbole at this stage, but it's possible that we could see a very different type of sporting contest emerge as the techniques available to coaches and players become more advanced as the years roll on. Returning to the example of tennis serving, it is already an established tactic to disguise the ball placement of the serve so as to deceive the returner of the likely delivery. Pete Sampras, seven-time Wimbledon champion, was famous for this asset in his famed service deliveries. As his regular foe Jim Courier has stated, '*the best servers don't give anything away. With the same toss and the same motion they can hit… out wide, into the body and then down the middle… the ultimate practitioner of that was Pete Sampras. He was by far the toughest server I came up against.*'[24]

Postural cue training has not been the only facet of visual sporting skill to be approached with video simulation training. Pattern recognition within a competitive scenario, such as a football match, has also been explored to see whether or not enhanced decision-making skills can be developed by using this technique. In a manner similar to the occlusion studies above, a scene of dynamic play is presented to an athlete, who then must choose which option they would favour as the next move; for example, pass, dribble left, dribble right, shoot, and so on.

The reasoning behind such an approach has been that the decisions made by an athlete can be broken down and analysed in a manner that could never be feasibly re-enacted in a live game setting; the pace of real time play would just be too fast. The information garnered could be used to better inform the player next time around. And findings from such experiments are showing some signs of promise with improvements in athletes making the 'correct' decision in the tactical dilemma scenarios they were presented with in later tests.[25]

The challenge that sport scientists currently face is quite what the level of transferability such training might bring. Are improvements simply observable on the tasks that these athletes complete in training, or do they actually lead to improvements in the field? The findings seem to point both ways,[26] with some noting evidence of transfer, while other studies suggest that transfer is limited. If it should be shown that transfer of training is limited, one question that might be important concerns just how relevant the mode of instruction is in imparting these skills. This is a fundamental question because if large portions of these perceptual and decision-making skills are learnt, yet transfer of training is negligible in more contrived training settings, a likely culprit is the method of instruction. Perhaps one learns these skills in a manner different to conventional coaching techniques.

Is preaching *really* teaching?

So, what is the best way to encourage the learning process? Is direct teaching, informing the athlete of the exact postural cue they should be looking for, the best learning approach? Or is a subtle, implicit approach a wiser strategy, allowing the player to discover the solution for themselves (with guidance where appropriate)?

One might assume that teaching a young boxer the precise mechanism of the uppercut, or exactly what needs to be observed in order to make a successful combination, would be the most fruitful of endeavours. If you neglect to inform the young hopeful of the crucial detail that divides the successful execution of a technique or tactic from one that is destined to fail, they may not notice this detail by themselves. And after all, the job of the coach is to tell the athlete how to play better. Important aspects of a technique often have to be transmitted overtly or else the vital components might become lost in translation. So there is certainly a case for direct teaching.

Nonetheless, research suggests that a more covert form of knowledge transfer can lead to greater gains in the long term of elite athletic development. The sentiment is that while an overt and prescriptive mode of teaching illuminates the exact factor the coach wishes to alert the athlete to, this method results in a less automatic ingraining of the skill in question. This is particularly evident in studies that show those who learnt skills in such a manner are more prone to choking under pressure and reverting to a very conscious mode of operation,[27] far detached from the instinctive plays of the elite sportsman. Conversely, implicit skill training has been shown to engender a deeper, subconscious grounding of the skill being learned and seems to promote less choking in athletes when the going gets tough.

With regard to learning elite perceptual and decision-making skills – abilities that need to be wholly automatic given the mere fractions of time typically available – this principle of implicit training would seem to be the route to follow. It would also seem likely that this is how these advanced skills have actually been imparted to those elite athletes who possess them. Indeed, the notion of improving an athlete's perceptual skills has not been considered an achievable possibility to many coaches until recently and so it seems unlikely that a formal and prescriptive training protocol would have existed. Coincidental training interventions have perhaps filled this crucial gap inadvertently.

One possible example of this occurring might be seen in the brilliant creativity of the Brazilian national football team. Their tremendous physical flair and precocious touch on the ball has been claimed to result from the lack of formal youth leagues in Brazil that might otherwise have stifled their creativity. As Leonardo, the World Cup-winning midfielder, has said, *'the children play a lot but it's always very free'*.[28] And could the physical skills that are developed in this way also be complemented by the emergence of elite perceptual skills?

As creative aesthetic football is so valued in Brazil developing young players are likely to be exposed to a greater (and richer) panorama of possible plays and strategies. This exposure would seem to mean that the young Brazilians are implicitly exposed to a large range of possible tactical plays – by virtue of their nation's obsession with beautiful football. Accordingly, they are implicitly guided towards a rich library of possibilities that is clearly represented in their style of play. The outcome of such skill development is that whereas a young English protégé might opt to simply run hard and fast at the opposition in a moment of tactical uncertainty, his young Brazilian counterpart might have something more unexpected in mind.

The lesson for coaches who are not fortunate enough to be working in the Brazilian beach clime is clear. Providing the opportunity for implicit learning of a broad range of patterns in the opposition play is a crucial component of acquiring advanced perception and decision-making skills. But how might this be achieved?

One possibility is to restrict the options open to young athletes – such as requiring an U/9 football team to only play with two touches – thus discouraging mindless runs with the ball. Instead, the players are forced into constantly evaluating the opposition's position and making evaluations of the best strategies to enact. The implicit nature of this set-up only serves to make this mode of learning deeper and more automatic and the ingenuity of the coach is perhaps the only limitation to implementing this approach in any number of other sports.

What? No genes!

The vast bulk of the chapter thus far has emphasised the fact that a magic eye existing for elite athletes is little more than a popular myth. In its place the role of learnt skills has been emphasised; specialist cognitive strategies that have developed in the mind of the best performers certainly seem to differentiate the great from the good. But we haven't reached the end of the tale just yet. Indeed, while many in recent times have become strong advocates of the role of nurture in sport, the interdependency of nature and nurture illustrated in chapter 2 has shown that the role of nurture, no matter how influential, can only be part of the story. Genes are a likely co-conspirator in the development of perceptual wizardry.

So, where does one look for the innate influence on the skills of perception and decision-making? At first this is a challenging proposition. If perceptual and decision-making skills are learnt abilities, then the crucial influences would seem to all lie in the realm of what exposure an individual has had to elite perceptual challenges. But this is to gloss over the fact that genes may well differentially affect the ways in which individuals are able to learn and utilise the visual knowledge they are exposed to.

One such influence is that people have very different visuospatial construction abilities.[29] That is to say, we all differ in the extent to which we can form visual images and manipulate these visual displays in the mind's eye. Unsurprisingly, those who are less able to render a complex representation of the outside world in their mind's eye are less likely to comprehend the emergence of complex patterns in a visual array and so will be at a disadvantage to those who can.

Importantly, this is not the same as better vision *per se* – the argument which was debunked in the opening sections of this chapter. This differential among individuals is a highly subtle cognitive feature, rather unlike the physiological abilities, such as better vision or faster tracking of moving objects, which were shown to be incompatible with the actual skills elite athlete hold over novices. And genes appear to hold an important portion of the variation in this visuospatial construction ability among the population. Indeed, the genetic influence on the variation in the population has been measured (through the twin study method) to be in the region of 44–68%.[30]

It is unlikely that the innate influence on perceptual and decision-making skills should end there. We have already seen in chapter 3 that the basic tendency of openness is highly influenced by one's genes and it would be surprising if those who were more open to new situations were not inadvertently exposed to new tactical plays that can then be stored in one's internal 'play scripts'. And further innate influences would also seem likely.

Drawing back to the interdependence of nature and nurture, perhaps the best way of viewing these different forms of influence on perceptual sporting skill is through the lens of the Matthew effect. One with great visuospatial abilities, coupled with an excellent learning environment is likely to enjoy success (with regard to the perceptual and decision-making aspects of sport) as these two features interact with each other. However, one without such fortuitous visuospatial abilities might still enjoy the sheer drama of a football match Brazil-style, but fail to register the finer nuances of the tactical play. As we have seen with the Matthew effect, small differences can quickly compound and influence other aspects of a young athlete's development. Indeed, one with early tactical awareness may be exposed to a more challenging environment, and so the cycle continues.

Conclusions

The overriding sentiments that I hope one is left with following this chapter are twofold. First, the notion of a magic eye in sport is a figment of the imagination, little more than an oft-convincing, and to some extent plausible, explanation for the seemingly miraculous perceptual and decision-making skills of the champions of sport. As we have seen, there are all-too-human underpinnings to these abilities. Second, these abilities do not appear to be the sole provision of an advantageously innate predisposition; rather, it seems likely that they can be learnt like any other skill that one is able to develop through extended practice. This is a profound change in perspective to the one seemingly held by many sports coaches who often refer to anticipation as if it is a kind of divine power.

However, this is not to say that innate influences have no effect in this domain. We have seen how individuals vary greatly in their visuospatial construction abilities and that these abilities have a sizeable genetic component. This of course points towards certain people having a more favourable predisposition to dissecting the sporting visual array than some others.

Nonetheless, the face of perceptual and decision-making skill training looks set for a potential coaching revolution. Video simulation equipment becomes more accessible by the year as digital cameras continue to plummet in price and portable projection screens are increasingly commonplace. Accordingly, it is likely that coaches at all levels of their respective sport will begin to incorporate these learning techniques into their coaching arsenal. This is an interesting projection into the future of coaching. Quite what this might mean for sporting contests is anyone's guess, but it would seem likely that if athletes can be trained to anticipate and predict their opponents' moves further ahead of time than previously managed, the opposition will be forced into new avenues of attack. Maybe the advent of such coaching techniques will, in their wake, yield unique, original techniques and tactics that will push the boundaries of sport ever closer to the limits of human capabilities. The next decade or so of sporting evolution promises to be fruitful in this regard.

It is now time to leave the domain of perceptual and decision-making sport skills, as we set our course towards another – the intriguing world of mental toughness in elite sport.

5

when the going gets tough, the tough get going

The annual Hawaii Ironman represents the pinnacle of worldwide triathlon competition. With its back-to-back 2.4-mile swim, 112-mile bike ride, and 26.2-mile run it is perhaps the most demanding sporting event in the world today.[1] And the legacy of the event owes a debt of gratitude to one triathlete in particular, the American Julie Moss. In 1982, while triathlons were still a relative anomaly on the world-sporting calendar, the 23-year-old Moss entered the Hawaii Ironman with the primary aim of gaining research material for her college thesis in exercise physiology.[2] She was certainly not expected to feature among the leading female triathletes of the day. Yet, several hours into the race, with just a few hundred yards left to run, Moss found herself doing the unthinkable. She was in first place and on the verge of gate crashing the elite triathlon community with a stunning and unexpected debut Ironman victory. However, Moss's legend has not endured because of a famous underdog victory. Rather, it was the events that unfolded in the very final stretch of the race that immortalised Moss and helped to establish the very ethos of the race, as captured in a short piece of television footage that can be described as nothing less than quite astonishing.[3]

An utterly fatigued Moss entered the final kilometre dangerously dehydrated and barely able to run to the finish line – '*the rest of the world slowed to a stop. I stopped seeing the spectators, the cameramen*'.[4] Inevitably, her legs gave way and she collapsed to the floor. She rose wearily to her feet but fell again shortly after; despite her numerous attempts to stand upright she simply could not muster the strength to do so. And amid this drama upon the homestretch, the second-placed triathlete, Kathleen McCartney, overtook her. Yet, despite her debilitating fatigue and the loss of first place Moss will be remembered for her stubborn refusal to receive help in making the finish line, aware that any assistance from stewards or the crowd would disqualify her from the competition. Instead, she crawled the final yards on her hands and knees, to the astonishment of the watching crowds who cheered her remarkable bravery to the line and beyond.

The courage of Julie Moss in those harrowing moments is one of the classic examples often given of an athlete displaying incredible mental toughness while under extreme duress. Indeed, one is instinctively aware of the athlete who is psychologically impervious – sportsmen such as Roger Federer, Tiger Woods, and Michael Jordan spring to mind. However, when pushed to provide a functional description of what mental toughness actually entails most people are somewhat unsure of what this concept means.

Mental toughness has often been likened to a ruthless coldheartedness or a killer instinct, such as that of the former world heavyweight-boxing champion Mike Tyson, who was feared throughout the boxing world as a vicious destroyer of opponents. Sheer stubbornness in the face of possible defeat is also routinely alluded to as a feature of the mentally tough athlete; Julie Moss's tribulations are a case in point. Yet one does not have to search very far to discover that these qualities are somewhat crude descriptions of the fortitude elite athletes' regularly show in the face of great competitive stress. Indeed, while a degree of ruthlessness will get one so far, athletes are known to perform better when they operate on positive sources of energy, rather than on negative vibes. And while stubbornness might well aid an athlete under certain circumstances – such as simply making it to the finish line – it also suggests a degree of mental inflexibility. The ability to strategise and adapt one's game plan, while absorbing a physical and/or psychological beating, is of course the hallmark of elite performers.

This, of course, is not to belittle Julie Moss's feats in any way; she was undoubtedly courageous and tough to the end. However, it would appear that there is more to the concept of mental toughness than often noted, and certainly more so than a stubborn refusal to give in, despite the many advantages of such fortitude. As in previous chapters, there is an explanatory gap of sorts that needs to be addressed.

So let us start with the seemingly obvious question, '*what is mental toughness*'? In short, mental toughness essentially entails, '*being able to cope with adversity in competitive situations*'.[5] However, this seemingly straightforward definition belies what is in fact a range of psychological skills and abilities, each packed inside this single concept. At least three essential components are central to the concept of mental toughness – (1) elite concentration, often amid great distractions, (2) a high capacity to cope with adversity and/or defeat, and (3) elite emotional regulation. Let us begin by looking at the underpinnings of the first of these components, the capacity to totally focus one's mind on the job at hand.

Keeping the mind on the job

Everyone knows what concentration is, if only because we have all experienced the loss of it at some crucial past moment. The ability to keep one's thoughts to the task at hand, in spite of distractions and temptations to the contrary is an important quality for all elite sportsmen. All too frequently one will hear of an athlete's defeat being explained away as the consequence of a *loss of focus* or *I let my mind wander.* Even tennis' king of clay Rafael Nadal admitted as much in a recent early round surprise loss, '*[I] started well today and took the first set playing fine tennis, but then I somehow lost my concentration, so it was a bit of a mental problem*'.[6]

Some have been more unfortunate, suffering despicable distractions to their performances, such as was seen in the 2004 Olympic marathon where a drunken spectator disgracefully forced race leader Vanderlei De Silva off the road. Although De Silva still went on to take the bronze medal he was understandably distracted from his race plan, being caught by the chasing pack shortly after the unfortunate altercation. '*I had to get back into my competitive rhythm, and I really lost a lot of it. It's extremely difficult to find that rhythm again.*'[7] There is no contest to the assertion that concentration is a valuable asset to any competitor, and despite the woes of Nadal and De Silva above, elite athletes are typically better at keeping their focus than the mere mortals of the sporting world.

Despite the relative clarity of the term, what actually defines the focus and concentration of an elite sportsman? What kinds of psychological processes underpin the ability to organise one's focus to the sports field, and not to the belligerent heckler in the third aisle of the bleachers? The concept of what psychologists' term *executive control*[8] is an important feature of one's concentration. This concept essentially refers to a set of cognitive abilities – such as inhibiting unwanted behaviour or dividing one's attention between tasks, among others – that collectively serve to guide thought and behaviour in a controlled manner.

One way of viewing such executive control in the brain is through the metaphor of the orchestral conductor.[9] The many different musical parts of the orchestra need guidance so as to integrate their unique abilities into a single coherent symphony. The role of executive control in the brain is to service this requirement of one's cognitive processes, just as the baton-swirling maestro will do with their musical charges. In fact, damage to the regions of the brain where such functions reside, the frontal lobes, often results in an individual who struggles – often with shocking disorganisation – to manage their goals and plans.[10] The question that begs to be asked at this point is whether it is possible to train one's attentional resources so as to keep one's mind on the job? Can one improve the processing power of the brain and thus enhance one's capacity to focus amid strenuous competition?

The high-street stores would probably argue that concentration (and other brain functions) can be improved through training, and literally scores of products have made their way to the marketplace purporting to aid the enhancement of such cognitive skills. From specialised computer games to mental quizzes, the list of devices purporting to improve these functions goes on and on.

Intuition might suggest that such tasks are likely to result in improvements in one's ability to filter out distractions, and so on, in light of the commonly held view that the brain – like a muscle – is amenable to training. Considerable evidence exists to support this view. For example, older adults who are mentally active – have challenging jobs, read regularly, and so on – are less likely to succumb to dementia.[11] However, the true effects of this attempt at cognitive enhancement are still unknown. Indeed, where improvements have been shown for a particular task, more often than not the improvements have remained in the domain of the task. That is to say, while one might enhance one's 'brain age' within the realm of a computer game, whether or not the improvements generalise to the underlying cognitive faculties that most would rather improve is unclear.[12]

Indeed, the significance of reported improvements in such forms of cognitive training has thus far been extremely minimal.[13]

The importance of concentration to the competitive sportsman is undoubted and so the suggestion that the 'hardware' underpinning concentration is less amenable to major improvements is disappointing. Nonetheless, what goes wrong when one's concentration disappears irretrievably? And is there potential for intervening strategies to be undertaken to aid one's focus, even if attempts to achieve 'pure' concentration improvements are as yet perhaps futile practices?

One intriguing theory has been suggested by the psychologist Daniel Wegner, who claims that the very attempt to control the thoughts in one's mind can actually cause one to lose focus.[14] After all, if one were told, *'don't think of purple penguins playing football'*, very often the harder one tries to ignore this thought the more difficult it becomes to do so. In a sporting context, if a heckling spectator has rattled an athlete's concentration, intuitive experience suggests that the more the athlete tells himself to ignore the heckler, the more he becomes aware of their presence. But how can this be? Haven't we been reliably informed by our coaches and teachers to block out such distractions so as to shield oneself from distraction? Is this strategy really such a futile one?

Wegner claims that control of one's thoughts is achieved through two interdependent processes: (1) intentional operating processes and (2) ironic monitoring processes. Intentional operating processes are those mechanisms engaged when one consciously attempts to attend to, or suppress, a particular feature of the environment. A Premiership footballer playing away in front of a volatile home crowd might say to himself, *'just ignore the jeers'*. But as one is already aware, this strategy does not always work. Wegner claims the reason for this is that ironic monitoring processes exist as an unconscious check that conscious mental control has not failed. In the example of the footballer, his monitoring processes will be working to make sure that if he entertains any thoughts inconsistent with those desired, they will be brought to conscious attention so he can deal with them appropriately. If he starts thinking of the hostile home crowd, prompts from his monitoring processes will allow him to initiate conscious control against the distraction.

While this unconscious monitoring system might sound useful to the competitive athlete, it may well be the reason for many of the frustrating lapses in concentration that are prone to occur at the most undesirable moments. Indeed, by attempting to suppress a particular thought, one inevitably primes the thought even more so. This is because one's mind needs to know what it shouldn't be focusing on, in order to not focus on it. This has led to Wegner calling his theory one of ironic processes; the harder one tries to ignore something, ironically, the more activated it becomes in one's subconscious and is more prone to bleed into one's conscious thoughts. This is particularly likely when individuals attempting thought suppression are tired or stressed – not an unusual state for an athlete – as the ability to perform conscious control may well be diminished.

With this thought in mind, the manner in which an athlete organises their focus might be more or less fruitful depending on whether they adopt a 'negative' stance, such as '*don't think about the pressure*', or a 'positive' stance, such as '*throw the right-hand punch when the guard slips low*'. Indeed, while attempting to suppress a distraction – such as the pressure one is under – one inevitably activates the monitoring processes that can remind one of the causes of the pressure. However, for the athlete who focuses upon performing a particular task, rather than on avoiding the distraction, the future appears to be bright for they do not create the monitoring feedback that can prove to be so distracting. The difference in outlook appears to be a profound one, and seems to have been adopted by elite athletes as a competitive tool. Olympic gold medal-winning sprinter Michael Johnson seems to be a fine example of an athlete who has not fallen foul to this trap, '*I have learned to cut out all the unnecessary thoughts … I concentrate on the tangible … on the race, on the blocks, on the things I have to do … and now it's just me and this one lane.*'[15]

Johnson is clearly illustrating that he concentrates best when he focuses on what he *should* be doing, rather than on *not* concentrating on the things he knows he shouldn't. The US Open champion golfer Michael Campbell has also spoken of how this form of positive concentration works for him, '*You need to stay in the present. I do this by focusing on something like a red shirt in the crowd or a really beautiful tree. That might sound funny but it makes you think [of] what is happening right now. Not what went before or what is going to happen.*'[16]

Sport psychologists have promoted a number of techniques that tap into and aid this mode of thought in recent times. One such approach is the use of pre-performance rituals. Such rituals might include bouncing the ball for a set number of times before serving, clapping one's gloves together in certain pattern prior to the need to make a penalty save, or by not walking on the lines of the court (a somewhat more superstitious habit). In fact, a sizeable proportion of elite athletes demonstrate some form of pre-performance ritual prior to executing an important activity or technique in their respective sports. The Serbian tennis champion Novak Djokovic frequently prepares for his services by bouncing the ball up to 20 times in a ritual he has claimed aids his concentration (others more sceptical have suggested it is little more than a sophisticated form of gamesmanship[17]).

There are a number of reasons why such rituals might aid an athlete's concentration. The first is that they act as a hard-wired reminder for the athlete exactly what it is they are about to do and what is required in order to successfully execute the task. For example, for a tennis player about to deliver a serve, bouncing the ball in a set pattern is likely to spark off automatic behaviour that overrides the temptation to 'over-think' the impending task. Another potential reason why these rituals might be of use is that, in line with Wegner's arguments, the intentions of the athlete are directed *towards*, and not away from, a goal. Thus, the likelihood that their internal monitoring processes will continually reactivate the distracting image is lessened.

And the use of routines has been shown to be a scientifically credible one for enhancing performance. In a study of two groups of amateur golfers,[18] the first group were taught just the technique of the swing, whereas the second group were taught the technique alongside a specific pre-shot routine that they were asked to use in the same order every time they played the stroke. The routine was, in essence, an arbitrary set-pattern – take two practice swings, glance up once at the hole, set up the grip, set the feet, glance up at the hole once more, swing. The pattern in itself was not believed to aid the learning of the stroke *per se*. However, the results of the study showed that the second group demonstrated greater improvements than the first group, a fact that was attributed to the benefits of pre-performance rituals in directing one's focus appropriately. This finding has been supported by the champion golfer Michael Campbell – '*more than anything you need to have your routine. I walk up to a shot and look where I want it to go. Next I take a practice swing. Then I cock my club back, once looking at the target, then two more times*'.[19]

Another approach claimed to improve concentration in sporting situations has been that of simulating competition stresses. Such an approach might involve getting a young and competitively inexperienced amateur boxer to spar with an opponent while the entire gym is made to watch the session and involve themselves with the proceedings (cheering vociferously, supporting the opponent, heckling, and so on). The Australian ladies hockey team used such a strategy to prepare for the 1988 Seoul Olympics by playing loud crowd noises over the tannoy in their training ground to simulate playing under competitive conditions – '*the players were ... asked to play and train with a public address system blearing out loud crowd noises by the side of the pitch*'.[20]

One might question how effective such a technique could ever hope to be. Indeed, what could ever truly prepare an athlete for the crucial penalty shot in a major championship final? The temptation to ponder on the implications of missing the shot – letting down one's nation – are simply out of proportion to those one will ever be able to replicate on the training ground. Nonetheless, playing the Koreans in the final, the home crowd was naturally against the Australian visitors. Yet the pressure of a hostile reception did not distract the team captain, Debbie Bowman, who struck a decisive penalty shot to set up the gold medal winning victory.

It is hard to conclude very much about the origins of Bowman's mental toughness from this one penalty. Accordingly, the effects of the simulated training are perhaps little more than an appealing anecdote at present. However, it would be surprising if such approaches were wholly redundant at all levels of the game. While one cannot truly recreate the effect of the hostile home crowd, for a young athlete one can surely use such a technique to introduce the higher levels of competition in controlled manner. A promising young footballer might be introduced to a more provocative match environment by playing with adults in a local Sunday league set-up, yet at a skill level where they are not overwhelmed. There is also surely something to be said for simulation training, if only as a placebo. If an athlete feels that the preparation they have received enables them to cope with the real-life version of the simulation, they are likely to adapt to the challenges with greater effectiveness.

Coping with adversity

So, one aspect of mental toughness – elite concentration – is an ability that one can improve through the use of elite strategies. However, despite the best attempts to direct one's focus and harness it to a competitive advantage, defeat – that oh-so-loathed word of all champions – is inevitable, even for the finest of competitors. Even the apparently invincible Mike Tyson came unstuck in the Tokyo Dome in 1990 against the 42-to-1 betting underdog James 'Buster' Douglas, in what has often been referred to as the biggest upset in the history of heavyweight championship fights. And one will not have to search too hard to find countless other examples of surprise defeats, although perhaps not quite in such dramatic fashion as the Douglas–Tyson encounter. Accordingly, while elite concentration is one facet of mental toughness, the ability to cope with adversity – impending defeat, crowd heckling, and moments of tumbling confidence – is an equally important aspect of mental toughness that sits alongside the capacity to focus on the job in hand.

Aspects of the self-efficacy theory of Albert Bandura were briefly described in Chapter 3 alluding to the notion that those who have experienced prior success are (unsurprisingly) more likely to possess a strong sense of self-esteem. Certainly, those who possess a greater confidence in their abilities, having seen them pay dividends in the past, would be most likely to see the guiding light in moments of great professional (and what often turns into personal) darkness. Those athletes with confidence simply oozing at the seams will be the ones who will best deal with a succession of adverse results and the challenges to their supremacy, right? Somewhat surprisingly, this sentiment has shown itself not to be the case, or at least not in the manner one might be expecting, as an insightful set of studies from the domain of educational psychology have shown.

We have already visited the concept of meaning-systems, the notion that we as humans organise our worlds around belief structures that we use so as to coherently navigate the ebbs and flows that life deals up. With this notion in mind, the psychologist Carol Dweck has argued that there exist two frameworks from within which an individual can understand the origins of their abilities and subsequent achievements:[21] (1) a theory of fixed intelligence and (2) a theory of malleable intelligence. The fixed intelligence theory refers to the sentiment that an individual holds the basic notion of his or her intelligence, or for our concerns – sporting ability, as something concrete and essentially fixed. As a result, such people are inclined to believe that their abilities are largely unchangeable. This kind of athlete might be characterised by thinking that their flaws (or indeed their strengths) are natural deficiencies; that they just don't have the kind of physical or psychological skills to be a successful athlete.

The reverse sentiment – the theory of malleable intelligence – would mean that an individual holds the belief that intelligence (again, read sporting ability) is a quality that can be altered for the better (and thus presumably for the worse as well). Such an athlete might well recognise that there are limitations in their abilities; one who is under a certain height is unlikely to play college or NBA basketball. However, the overriding sentiments held by this athlete are that sporting abilities are developed through training and that they fundamentally emerge following good, old-fashioned practice. The NBA Hall of Famer Michael Jordan echoes such sentiments when commenting that '*nothing of value comes without hard work*'.[22]

The logical consequences of such a way of viewing one's capabilities are somewhat profound. For the malleable theorist a challenge to one's success, while undoubtedly an irritant, can be appraised and understood by the individual as a need to work harder, make appropriate changes to one's approach, and so on. Not so for the fixed theorist. This individual will likely appraise his or her actions somewhat differently. For this individual a knock-back or a loss is a blow to the self-esteem unlike that experienced by the 'malleable' theorist because as their ability is viewed as a static entity there seems to be little they can do about it. And so, while these individuals are having their dreams of victory dashed – a disaster for any athlete – they also face having their very ability as an athlete (as they perceive it) exposed to all and sundry as limited and vulnerable. They cannot hide behind the notion that they can return to fight another day with more practice and better strategy behind them.

In support of such a contention, a series of studies by Dweck has shown that those who display the characteristics of the theory of fixed intelligence are more likely to show signs of psychological 'get-outs' when pressure and/or failure inevitably occurs.[23] In a sporting context this would relate to behaviours such as excuse making, feigning disinterest, and/or claiming that the opposition were cheating. The purpose of such a ploy would be to alleviate the necessity for the athlete to take the burden of failure squarely upon their own shoulders. It allows those who hold theories of fixed ability to not have to take the crushing news that they were just not good enough, and that there is nothing they can do about it, announced to the sporting world.

Perhaps this is a sound modus operandi. After all, who wants to have their vulnerabilities exposed in such a manner, especially in the cruel pantheon of junior sports? But while this might make good sense in the heat of the moment (after all, saving face is a major preoccupation of many young athletes), a more serious implication of this strategy is likely to rear its ugly head. Indeed, those who make a serial habit of 'tanking', for fear of a vulnerable loss, will inevitably be those individuals who, often despite great physical promise, are unable to muster the competitive skills that allow challenging moments to be overcome. They simply will not be mentally tough when it matters most. So, what are the origins of these theories, which hold such important implications to any athlete's ability to cope with competitive stress?

Dweck has suggested that the manner in which others – especially influential adults such as parents, coaches, and teachers – appraise the successes and failures of an individual plays a crucial role.[24] Those who are consistently told how great they are, and what talent they possess, are provided with the basis for which to develop the mindset that underpins a fixed theory of ability. In essence, they are being groomed to feel that the sporting ability they enjoy is not the product of anything one has strived for, but instead is the product of natural giftedness. The nature–nurture debate framed in chapter 1 has shown just how prevalent this notion of innate giftedness often is, and so it is not surprising that young athletes buy into this mindset. And this seems to hold true when one considers the classic syndrome of the hyper-talented junior athlete. How often does one see the super-gifted young athlete wallow in fits of rage and partake in the tanking of matches or competitions? Perhaps the years of being informed of how 'special' one is gradually lead a young athlete towards fearing the loss so much so that they are no longer able to perform when the pressure is on. After all, who wants to come to the conclusion that they are really not so 'special after all?

On the other hand, those who are reminded that hard work is the crucial link to success are more likely to develop a theory of malleable ability. The groundwork will have been laid confirming to the individual that the achievements they have earned were made through a diligent and persevering approach: '*well done son, you really made all the hard work pay off today!*'

But there is surely some kind of trade off. Indeed, it would seem likely that a combination of the two facets in specific doses might be more effective than simply ascribing to a malleable theory of ability. For the young athlete who deeply believes they are physically gifted (and may well be), yet is only too aware of the importance of hard work in cultivating their gifts, the sky is perhaps the limit. Nonetheless, Dweck's work has enormous implications for concepts of how mental toughness needs to be viewed and it will be interesting to see how coaches and educators take up such concepts in the years to come.

Getting emotional

Anxiety is often seen as a detrimental aspect in any athlete's quest for a rousing sporting performance on the big occasion. How many suffer a poor night's sleep or need to visit the bathroom repeatedly before an important match? Despite his calm exterior England rugby hero Jonny Wilkinson has felt the very same way, '*there are times where... shooting out the back door, hopping on a plane and getting a quick holiday for two weeks... [before a big match] It's the sort of thing going through your mind because it's quite painful*'.[25] But emotional energy is also important for an athlete. After all, without a certain degree of residual fear (of defeat, embarrassment, loss of a team place) the adrenaline that propels an athlete to the top of their game simply would not exist. All of those who compete, or have done so with any regularity, will know that familiar rush of adrenaline through the veins that spurs one onto the finish line, even when all else seems to be lost.

These surges of adrenaline in times of stress are a response to danger that has evolutionary underpinnings, and are derived from aspects of life far deeper-rooted in humans than mere sporting contests. Indeed, if a grazing zebra should catch sight of a roaming predator, such as a hungry lion, what is known as the *fight-or-flight response*[26] will become activated in the zebra. From a resting state the zebra will suddenly find its physiology changing dramatically. Blood vessels will start constricting, pupils will dilate, and intestinal actions are halted; all with the explicit aim of preparing the animal for an exceptional burst of energy so as to make its escape from the fatal jaws of the prowling lion.

And so it is with athletes facing a great sporting challenge. The rush of adrenalin prompted by the significance of the big-match occasion prepares the body to deliver an energised performance. In fact, for this very reason the greatest performances often occur when sportsmen go toe-to-toe against their fiercest rivals. Conversely, many talk of feeling flat and turn in below-par performances when competing against those who don't prompt any emotionally arousing sentiments. The world boxing champion Joe Calzaghe has echoed both sentiments in a number of performances throughout his outstanding fight career. His poorest performances have often come against sub-standard opponents, whereas his greatest victories have arisen from the cauldron of self-doubt and fear. Indeed, former world champion Nigel Benn has commented, *'if he fights world-class opponents he looks good, but when he fights lower level opposition he is brought down to their level'*.[27] And as American track superstar Michael Johnson contemplated shortly before his historic sprint double – winning gold medals in both the 200 m and 400 m at the 1996 Atlanta Olympics – *'I have never felt pressure like this in my life... but I run well when I am scared'*.[28] Certainly, no one who saw him race in those events would doubt it. He decimated one of the longest standing world records in the 200 m

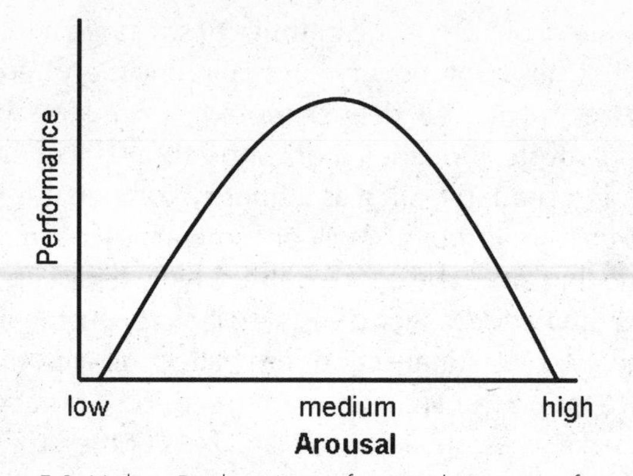

Figure 5.1 Yerkes–Dodson Law of arousal versus performance.

final, such that some in the stands were led to question whether or not the trackside timer had broken during the course of the race.

The proverbial butterflies one will feel in the pit of the stomach are thus a very necessary component of performing at the upper end of one's competitive abilities. This principle is illustrated with the Yerkes–Dodson law (see Figure 5.1), where the level of arousal is positively related to the level of performance. As one becomes more aroused, so the performance level is raised.

However, as the graph illustrates, if one's levels of arousal become *too* high, the reverse effect becomes apparent; the level of performance drops as one becomes gradually more and more incapacitated by fear and nerves. Anyone who has experienced competition will have felt the near-paralysing fear that can sometimes occur in the moments before one walks out onto the court or sports field in front of a watching crowd (be it one's work mates – or for those fortunate few, the millions who tune into Wimbledon each year).

With the example of Yerkes–Dodson law it will probably be no real surprise to hear that one of the great abilities of any mentally tough champion is the sense of when and how to control their emotional energies. The ability to produce the match-winning rise in intensity at the moment of greatest need, or to shrug one's shoulders and stay calm in the midst of turgid encounter is crucial on the elite stages of sport. Indeed, despite his reports of terrible pre-match nerves, Jonny Wilkinson has clearly managed to contain his inner demons and lead his country to a number of famous victories, not least the 2003 Rugby World Cup triumph, where his breathtaking winning drop-goal in the final seconds of the match broke the heart of the Australian team.[29]

But the ability to manage one's emotional state to cue is now not believed to be an innate skill that elite athletes possess over mere mortals of the amateur world (although the involvement of innate factors is discussed later in this chapter). Rather, it is an ability thought to arise following a host of conscious strategic ploys that affect an athlete's emotional states, deployed as and when they are required. That is to say, elite athletes have acquired a set of mental skills, which allow them to modulate their levels of arousal and thus compete with greater potency. With this thought in mind, what kinds of psychological strategies are being used by elite athletes in order to manage their emotions so effectively?

One such strategy is that of *visual imagery*.[30] Many of us can conjure a rich panorama in the mind's eye of any type of scene that one might care to imagine. Whether it might be the wet and muddy turf of a furious British Lions-All Blacks encounter, or the adrenaline-fuelled ring entrance to a world heavyweight title fight, each of us has likely been to an imaginary sporting world far from reality, yet rich in colour and intensity. And anyone who has imagined themselves in such a situation might well have felt a tingle run down the spine. Visual imagery can powerfully alter one's bodily states, and this effect has been shown in a number of studies that illustrate that imagining an image or scene utilises the same brain regions that would be activated if the scene were available for direct inspection.[31] The upshot of such findings is that visual imagery can be thought of as a tool to 'transport' oneself to a scenario that fosters positive sentiments – the cusp of a crucial victory, a soothing beach – and so altering one's bodily states accordingly.

Unsurprisingly, elite athletes have been observed to use specific imagery in order to provoke bodily states that enhance their physiological and psychological arousal as and when required. Australian tennis star of the 1970s and 80s and former world number one Evonne Goolagong in describing her pre-match nerves has stated, '*just... closing my eyes and imagining myself playing well... helped me psych myself up*'.[32] The world boxing champion Oscar De La Hoya famously dedicated his 1992 Olympic victory to the memory of his late mother Cecilia, who died of breast cancer the year before the Barcelona Games. It also appears that the death of his mother provided Oscar with a powerful visual image, that of winning the gold medal for her as it had been her dream as much as his; '*one day I said to myself, "if my mother was alive she would want me to continue on and go for the dream"... and she did tell me to go and do it*'.[33] And as Muhammad Ali has stated with typical panache, '*the man without imagination has no wings, he cannot fly*',[34] alerting one to the power of seeing a great performance happening in the mind's eye before the first bell sounds. And visual imagery, as well being used to spark high levels of arousal, can also be used to calm an athlete and reinstall a sense of togetherness between mind and body. The motor racing icon Jackie Stewart was known to utilise a visual imagery routine in the minutes leading up to races. '*He would sit in the car moments before the start and imagine his body inflating like a balloon. Then he would let the air out and feel himself relax. This, he contended, helped him prepare physically and mentally for a race.*'[35]

In addition to visual imagery, self-talk is another strategy that athletes in stressful situations are known to utilise in order to regulate their levels of arousal. Self-talk is that familiar dialogue we each partake in with ourselves – either internally or out loud – and it has become clear that the words one uses in this ongoing narrative influence one's psychological states. Indeed, studies[36] have shown how athletes who exhibit positive self-talk prior to an event – '*you know this is gonna be your day*' – are more likely to perform with greater success. And this does not simply seem to be the result of the better athletes already knowing they are the best and thinking accordingly. In studies of impulsive children, those who were given a programme of cognitive self-guidance, which included advice on how to use self-talk, were subsequently better able to manage their behaviour.[37]

Accordingly, elite athletes have become increasingly aware of this way to modulate their levels of emotional arousal. It is commonplace to hear an athlete gee themselves up on the sports field with a fierce '*c'mon*' or similar cries. Muhammad Ali was perhaps the greatest exponent of this technique, routinely declaring himself to be 'the greatest of all time'. Perhaps as importantly, Ali got into the mind of his opponents like few before or since. Before his heavyweight title fight with Sonny Liston, Ali (then still Cassius Clay) repeatedly referred to Liston as the '*big, ugly bear*',[38] even driving to his home late one night to taunt him, all in an attempt to undermine Liston's confidence. Who knows what Liston truly made of Ali's antics but there is at least a chance he experienced a degree of self-doubt creep into the back of his mind, which may have prompted his controversial sixth round retirement in their first fight. England cricketing great Fred Trueman was known to try the same trick, although perhaps somewhat more elegantly. Trueman remarked to an incoming Australian batsman, who reached out to close the gate behind him, '*don't bother son, you won't be out there long enough'*.[39] The art of trash talking has clearly been around for sometime.

Breathing techniques are also often used by athletes to control their emotional arousal, although most frequently in order to calm themselves down. Breathing exercises often vary from individual to individual, but many take similar forms. Typically the athlete in question will stand, sit, or lie in a quiet room or a secluded part of the sports-field, and simply take a number of deep, slow inhalations and exhalations, all the while focusing on the sensations that the breathing process registers within the body. Such a technique, once well learned, can even be utilised within competitive action, such as before a penalty kick, or at the change-of-ends in a tennis match. And like physical skills they require practice, diminishing over time if they are taken for granted. The 2005 US Open golf champion Michael Campbell has said that in preparation for a shot he uses simple breathing strategies to control his emotions – *'I ... take a couple of deep breaths to get rid of all the negative energy and breathe in the positive stuff'*.[40]

While visual imagery, self-talk, and breathing techniques are useful internal modulators of energy levels, external sources of arousal are also often utilised by athletes. Music is one very well-documented medium with which to affect one's arousal levels. Thus, similarly to their imagery routines, elite athletes often have an agenda with their musical selection – they frequently have a handful of songs easily accessible via an MP3 player or an iPod that can be used in the moments before, or even during an event. British tennis star Andy Murray famously competed in his first Wimbledon with the sound of the Proclaimer's 1980s hit song *I'm Gonna Be (500 miles)* blaring out of his earphones at the change of ends, so as to psych himself up when he felt he needed the lift between games. The Premiership football team Manchester United have been reported to have been inspired in the dressing room by the rousing Iggy Pop classic *Lust for Life*,[41] probably better known to the players as the soundtrack from the cult movie *Trainspotting*. And there are many other such sources of inspiration or relaxation; movies, favourite motivational quotes, and so on.

Given that the ability to maintain emotional control has been shown to arise from a range of cognitive ploys – imagery, self-talk, and so on – one important question concerns where and how these skills are learnt? Until fairly recently the answer to that question might well have been put down to sheer luck. Before the advent of sport psychology – where such techniques are now formally taught – athletes might have discovered the effectiveness of such strategies by accident, yet have been wise enough to keep on using them, even if the science behind the ploy was unknown. Oscar De La Hoya might not have known the science of visual imagery, and how it could help his performances. But the thoughts of his late mother deriving pride from his Olympic results would have been likely to have triggered powerful resources of energy and it is possible that scenarios such as this implicitly found their way into his pre-fight psych-up routine.

In addition, it is probable that folk-psychology – wisdom of the mind's workings accrued from experience – will have been transmitted through the generations, even if in absence of sport science. Coaches and athletes might not have known of the reasons why imagery prompts changes in arousal levels, but that of course would not have precluded its use. However, this approach is also sure to have launched countless strategies of little competitive use. One such ploy that now seems somewhat incredulous is that of boxers believing that crash dieting to make the stringent weight limit of their fighting division instils hunger and desire.[42] It certainly would instil hunger – the literal early signs of starvation. Unfortunately, it is now well known that this hunger will impair performance, not enhance it, as the body (and mind) is drained of resources with which to function properly. The current role of the contemporary sport psychologist seems to involve rectifying such entrenched beliefs as much as accentuating the strategies that do work in managing emotional control.

Innate toughness?

As the previous sections have intimated, a large portion of current sports psychological thinking suggests that mental toughness is a nurtured quality; something one can acquire like any other skill. The evidence for this perspective is strong and many of the arguments for this position have been described above. However, mental toughness is not solely a learnable cognitive skill in the vein of tying one's laces or driving a car. One intriguing finding that has illustrated the role of one's inherited qualities leads one back to the personality literature. Hans Eysenck, one of the founders of personality psychology, suggested that great individual differences exist in the amount of stimulation filtered through to the brain for processing. With this in mind, Eysenck has made the counterintuitive statement that introverts are characterized by higher levels of cortical activity than extraverts and so are chronically more aroused.[43] As such, introverts are argued to be so inclined because they are already highly innately aroused. In other words, they are less in need of stimulation to keep their levels of arousal at a comfortable point. The reverse is argued to be true for extraverts. Accordingly, introverts might feel sufficiently positively aroused while watching an action movie, whereas extraverts might need to jump out of a plane to achieve the same level of excitement.

The relevance to the mental toughness debate concerns the claim that introverts should be less effective at performing tasks when placed in situations of considerable stress. The basis for this view is that as they are already highly aroused further stress could push them beyond their comfort zone and disrupt their ability to concentrate. One piece of evidence supporting this claim has come from a study showing that introverts were noticeably more distracted and fatigued than extroverts while performing basic arithmetic sums in a room of loud noise (recorded traffic sounds).[44] The natural question at this point is whether or not introverted athletes are equally prone to mental 'overload'. Unfortunately, such observations are not evident in the literature at this point in time.

Just as individuals have been claimed to show striking differences in arousal levels, the same is also true of executive control. To refresh, executive control refers to the abilities to manipulate attention, which are likely to be central to one's concentration. However, whereas some of us are very effective at manipulating our thoughts – for example, making a calculation in the mind's eye – others are less so. Intriguingly, recent research has suggested that these differences are highly heritable.[45] As such, one's genetic legacy may well play a role in one's ability to think straight during a tough sporting encounter. Again, this is a new area of research and so one will have to wait some time for a greater understanding of the genetic involvement in one's executive faculties but the promise for findings of real interest to sport psychology are likely.

Conclusion

It is clear, I hope, that the mysterious quality of mental toughness is not simply a callous and cold-hearted approach to one's sporting goals. Similarly, a hard-nosed and stubborn refusal to give in to the opposition is in some way wide of the mark of what is truly embodied by the notion of mental toughness.

That is not to say that these features are not present in some way, but merely that they are small aspects of what is undoubtedly a wider set of skills and abilities.

More poignantly, it has been claimed that mental toughness is not the birth right of a selected few. Rather, skills that facilitate one's ability to cope in times of stress are learnable, although there do appear to exist certain natural tendencies that allow some to manage better than others under competitive duress.

With this outlining of mental toughness, we have now seen the final aspect of the psychological armoury with which elite athletes are typically endowed. However, the route to a championship mind does not end here. Indeed, while the previous chapters have shown many of the factors that conspire to produce elite sporting psychology, one cannot ignore the fact that careful nurturing of the 'talent' is central to maximising the potential of a young athlete. And so in chapter 6, 'The Hothouse Kids', we turn to the dilemma of how to provide intensive elite training programmes for gifted young athletes without the unwelcome arrival of 'burnout'.

6

the hothouse kids

Specialist programmes of training the young can be traced back at least as far as the sixth or seventh century BC with the infamous example of the Spartan *agoge*.[1] This rigorous regime of training and education was utilised by the Spartan community as a means of preparing their young males to defend the city of Sparta against potential invaders. For a city constructed without walls, and thus vulnerable to attack, these young apprentice-soldiers were nurtured so as to become the living 'walls' of the city.

In order to ensure that the Spartan men would be suitably hardened and capable of vanquishing any foe they might need to face in combat, Spartan boys were removed from their families at the tender age of just seven years. From herein they were placed in the hands of the state and henceforth lived in dormitories with the other boys of their age. From this point on, until their military retirement at the age of 60, they became an instrument of the Spartan state. The boys were educated in the manner that befit their Athenian peers, receiving lessons in reading, writing, and music. However, the physical exertions to which these young Spartans were expected to adhere have placed their name in legend.

Boys roamed barefoot, in the lightest and flimsiest of clothing, regardless of the weather conditions. The reeds of the River Eurotas provided bedding and were cut and collected with bare hands. The food rations provided for the boys were minimal and so the young Spartans were deliberately kept from enjoying the pleasures of a full stomach. Furthermore, they were encouraged to steal so as to make up the nutritional shortfall. If caught, the boys were punished; but rather than this being for the offence of stealing, instead it was administered for the crime of getting caught. And so, through these infamously 'Spartan' measures, these young apprentice soldiers were thought to be acquiring the key skills needed to develop into cunning, physically powerful, and self-sufficient adults capable of securing the future of Sparta, whatever pleasures such a future might be thought to bring.

The rise of professionalism in sport

The world of sport is no direct comparison to the world that the Spartans inhabited all those centuries ago. After all, the cost of losing a match of football or tennis is in most cases little more than a blow to one's pride or a knock to one's self-confidence. A loss at the event for which these Spartan boys were prepared was likely to amount to a scenario somewhat bloodier. Yet the latter portions of the twentieth century have seen a change in the approach taken towards the training of young athletes, one that often approaches those of the ancient Spartan ideals. The transition of sport from amateurism to professionalism, a change that has been evident since the early 1970s, has almost wholly redefined the landscape of elite sports development.

To contrast the changes sport has made in the latter portions of the twentieth century, one only need reflect upon the rousing words of Pierre de Coubertin, founder of the modern Olympics in 1896 – '*The most important thing in the Olympic Games is not winning but taking part; the essential thing in life is not conquering but fighting well.'*[2] Baron de Coubertin's words epitomise an ancient sporting ideal, symbolising the time-honoured traditions of sportsmanship and honour. Certainly, in Britain prior to the 1970s, the notion of conducting oneself in a manner anything other than that of a gentleman was deemed a serious social *faux pas*.

But the dignified motto of de Coubertin, for better or worse, has now been largely eroded. The sentiments of those such as the legendary American football coach Vince Lombardi, when he proclaimed, '*If it doesn't matter who wins or loses, then why do they keep score?*',[3] now epitomise much of modern thinking towards sport. As we approach the second decade of the twenty-first century, it seems increasingly the case that the sentiments of Lombardi have never been more apparent. Indeed, the slogan of the sporting brand Nike prior to the 1996 Atlanta Olympics read, '*you don't win the silver; you lose the gold*'.[4] This is not to say that athletes of generations gone-by were not highly competitive but simply that the ultra-competitive sporting world is now an acceptable, and highly pervading, feature of modern society. Times have certainly changed and de Coubertin might well be rolling in his grave if he were to know of the erosion of his Olympic sporting ideals.

It is perhaps not surprising that this ultra-competitive age of sport is upon us. Athletes have always been heroes to the sport-loving public and many figures of the past and present have transcended sport entirely to become household names in their own right. Figures such as Muhammad Ali and David Beckham are two who immediately spring to mind. Each one has dominated news headlines across the globe and is well known to those for whom sport is little more than an afterthought in life. It is possible that the nation's teenage girls know more of Beckham's life than many ardent Manchester United fans did while he was their star attraction. And of course the regular heroes of sport command sizeable fan support in their own right.

This social phenomenon, coupled with the opening up of major sport events to commercial sponsors following the rise of professionalism, was surely always destined to result in vast sums of money being injected into the sports industry. The lure for the major brands to present their products to a mass marketplace was always going to be irresistible. The result has been that kings and queens of modern sport are the recipients of what might have been called 'new money' in earlier generations, and quite extraordinary sums of money at that. The reported £128 million[5] that David Beckham was reported to be earning in salary and endorsements following his transfer to the California-based Los Angeles Galaxy football club is staggering. However, even this sum pales when one talks of the golfer Tiger Woods who is said to become the first billion-dollar athlete.[6]

For the mere mortals of the professional game, while the figures are somewhat less glamorous, the financial rewards are certainly not to be derided. In the United States the average salaries for athletes in all of the major four sports – basketball, baseball, American football, and ice hockey – each amount to sums well into the seven-figure bracket. Many, of course, are earning several times this amount. Even footballers in the English Championship and League One (the divisions one and two places below the Premiership) command salaries averaging £200,000 and £70,000,[7] respectively. Underpinning all of this sporting wealth is a global sports industry that is worth $120 billion a year in the United States alone. These kinds of sums rank sport as a major player in any conversation of global commercial muscle.

Money makes the world go round

In light of these astronomical financial payoffs the pressures for a young prospect to succeed in sport are bestowed with all the weight of expectation that only the prospect of millions of pounds can bring. Stories that have made headlines in recent years include one British football club who offered the parents of a precocious five-year-old player £10,000 to win his signature for the club.[8] This is not wholly anomalous. Boys as young as ten years old being offered terms to sign for major professional football teams have been periodically reported in a number of European countries. Lewis Hamilton, who has since developed into one of the world's leading Formula One racing drivers, was himself contracted to McLaren at the age of 15 for the princely sum of $2 million.[9] Of course, with such financial imperatives the possibility for foul play is perhaps inevitable.

In 2006 one such case of foul play was splashed around the world's media and emphasised the changing state of world sport with regard to the importance of victory. In the early 2000s Christophe Fauviau's two children were promising young tennis players. His daughter was a special talent and tipped to make a statement on the professional tour. However, in 2006 Fauviau stood trial accused of drugging his children's opponents in order to secure their victories.[10] The sequence of events leading to the charge came after a 25-year-old opponent of Fauviau's teenage son withdrew halfway through a match in a local tournament suffering from nausea and dizziness. This in itself perhaps was not an exceptional occurrence on a warm summer's day. After all, the rigours of competition are likely to claim numerous scalps in this manner from time to time, if only from sunstroke and dehydration. However, the events that unfolded shortly afterwards were tragically exceptional. Following his withdrawal, Alexandre Lagardère set out on the drive home through a mountainous region of France where he lost control of his car after becoming drowsy at the wheel. The ensuing crash claimed his life. The resulting autopsy discovered traces of the sedative Temesta in Lagardère's system and the alarm bells were sounded. The finger of suspicion did not take long to reach Fauviau. Indeed, reports of other opponents, one as young as 11, collapsing on court either during or following encounters with his children hinted towards his foul play. Fauviau duly received a substantial prison sentence following his confession to the crime which led to such an unfortunate, albeit unintended, outcome.

A similar story of the obsession for victory, fortunately with a somewhat less tragic consequence than just described, concerned the figure skater Tonya Harding's role in the infamous assault upon fellow skater Nancy Kerrigan.[11] In early 1994 both ladies were among the elite skaters vying for a place on the United States Winter Olympics team due to compete in the Lillehammer games later that year. However, following a practice session prior to the United States skating championships, where selections to the team were to be made, a mystery assailant struck Kerrigan around the knee with a baton. The injury she suffered from the attack resulted in her withdrawal from the competition, an event which Harding went on to win and thus securing her team selection. But it was soon uncovered that the attack on Kerrigan was a premeditated assault arranged by Harding and committed by her husband so as to remove her strongest rival from contention. (A plan that backfired, for the selectors deemed that as Kerrigan's injury did not occur on the ice she should still be considered for a team place and was in fact duly selected.)

Extraordinarily, the media soap opera that ensued following the assault allegations occurred during the Olympic competition where both Harding and Kerrigan were competing, living, and training side-by-side in the athletes' quarters. In the end

Harding was charged and convicted for her role in the assault. Kerrigan almost wrote her own Hollywood ending coming within a whisker of the gold medal, eventually having to settle for silver, but vindicating herself in the face of her disgraced rival nonetheless.

Hothouse culture

What the Fauviau and Harding cases exemplify above all is the extraordinary lengths that individuals will go to in order to maximise the chance of sporting success for themselves or that of their charges. And while the extreme nature of these criminal events is largely anomalous, it is clear that a more systematic trend is occurring in youth sports following the advent of the global commercial interests in sport. The rise of 'hothousing' – the practice of placing a young protégé into an intensive training environment with the explicit goal of nurturing an elite performer – is on the increase. It is not unusual these days for young boys and girls who are perceived by coaches (or parents) as possessing the raw physical ingredients to succeed in sport to begin systematic and rigorous training, often for several hours each day. Martha Karolyi, the gymnastics coach to many Olympic medallists, has noted that '*With the little kids ... you might have two hours ... at the beginning ... and then go on to three hours at one of the workouts and then up to three and one half hours in the morning. Nowadays, to be able to compete at an elite level, it takes about seven hours of training a day.*'[12]

Science in recent decades has increasingly supported this extensive training legacy on the future prospects of an elite hopeful. Research now shows with striking consistency that top-flight performers, almost without exception, have accumulated something to the order of 10,000 hours of deliberate practice (systematic and specialised training) prior to their professional successes.[13] This has been noted in pursuits as broad as music, science, chess, and sport and has lent great support to the hothouse movement, which now believes (with some justification) that the ability to 'produce' an elite performer is possible given the right grooming.

The fact that training is vital in order to exploit any genetic potential is of course not in dispute here. But what is of interest is the observation that *such* extensive training has shown itself to be the crucial catalyst in transforming a young hopeful with the innate potential into a world-beater. What this amounts to in practice is that great feats are preceeded by close to four hours of training per day over the course of a decade (not including weekends by this calculation). And this training must, of course, take place while the promising young athlete is also completing his or her education, considerably adding to the load placed upon the shoulders of such individuals. Furthermore, such commitments are needed while the young athlete is still emotionally and psychologically immature, and it does not take a startling imagination to conclude that the challenges of this lifestyle are immense.

The Russian starlet Anna Kournikova is one well-noted hothouse kid who experienced a childhood devoted to achieving her tennis dreams. At the age of 11, Kournikova left her homeland, father, and friends behind in her native Moscow when she relocated with her mother to live and train at the famed Bollettieri academy in the United States.[14] Here, the young Kournikova was surrounded by dozens of other hopefuls who spent many hours on court honing the skills with which to make an assault on the professional ranks. It would not be uncommon for players in these environments to partake in up to six hours of physical training per day. One promising young British junior player highlighted the intensity of the American tennis hothouses – '*At home I get up at 7 am…At the academy we started our morning sessions at 5.30 am, which meant getting up at 5 am. By the weekend, when we got up at 6.30 am, it felt like a lie-in.*'[15]

Similar training regimens have also been observed in the gymnastics world. US Olympic star Mary Lou Retton has spoken of the intense training regimen that propelled her to Olympic gold – '*Two years prior to the Olympics, our daily schedule was 7:00 to 11:00 in the gym every morning. We would shower at the gym, go to school for a few hours and then back to the gym from 5:00 p.m. to 9:00 p.m. every single day. So it was eight hours every day.*'[16]

Burnout at the hothouse

Aristotle was perhaps the first to stress the sentiments of too much, too young. He noted of the ancient Olympic Games that '*there are only two or three cases of the same person having won in the men's events who had previously won in the boy's [events]*'.[17] Aristotle believed that, '*early training... had resulted in the loss of strength*'.[18] It is probably not wise to base the principles of modern sport science upon the wisdom of Aristotle. For all of their cultural and scientific contributions, understandings of the mind and body have moved on somewhat since the time of the ancient Greeks. Yet Aristotle was again ahead of his time. The very real danger exists that the young athlete who trains as if he or she is already at the physical and emotional maturity of an adult professional is destined to become a washed up, burnt-out adolescent.

In professional tennis it has been noted that the number of elite ladies who have retired from the tour at an age when they should be comfortably reaching their prime has risen notably. Martina Hingis retired at the age of 21 with chronic injuries amid general apathy; she made an ill-fated comeback in 2006 – she tested positive for a banned substance, cocaine – and returned to retirement somewhat ignominiously.[19] More recently, the Belgian tennis star Justine Henin, at the time the world's number one player, was another high-flying casualty of the rigours of professional sport. Her coach, Carlos Rodriguez, observed that the '*fire that drove her to success*'[20] had seemingly been extinguished, perhaps through the exhausting schedules the professional players need to keep in order to maintain their ranking.

In the men's game the enormous hype that followed America's junior phenomenon Donald Young has, in the eyes of some, irrevocably harmed his transition from world junior champion to senior grand slam titlist. Whereas most players of his age were still plying their trade on the less-glamorous futures circuit (professional tennis' bottom rung), Young was placed in main draws of main tour events by agents eager to expose his extraordinary talents to potential sponsors. He suffered many heavy losses as a consequence. As the United States Davis Cup captain Patrick McEnroe commented, '*you can't let a kid go out there and lose [heavily] 10 times in a row... It can start to have a detrimental effect on his mental state*'.[21] Young himself was quoted as having questioned his abilities following such losses, '*I felt like I wasn't good enough, felt like I should go to school now, just hang it up... I didn't feel I deserved to be in the locker room.*'[22] And although Young has recently begun to establish himself on the main tour, only time will tell if his career is as celebrated as his arrival.

Little is more discouraging than to observe a gifted teen with nothing to offer his sport except a sad tale of apathy and bitterness following an overdose of training. And so, wannabe professional athletes face a tricky dilemma. Clearly, they must put in the amount of hours that research shows is so crucial for converting budding aspirations into something more concrete.

But in doing so they run the risk of burnout. With this in mind, a fundamental aspect of the making of any sporting champion will be to successfully negotiate this tricky relationship. Accordingly, what is burnout and how can one prevent or alleviate its symptoms?

What is burnout?

Like many other concepts in pop-psychology, burnout is not a term that has proved to be wholly understandable, at least when referred to as a definitive unitary complaint. Essentially, it describes the mental unrest of an athlete who has become psychologically and emotionally fatigued with training and competition.[23] The characteristics of burnout are many; however, certain features are most typical.[24] Difficulties in sleeping, unusual irritability, prolonged tiredness, and a poor appetite are all signs of psychological staleness and burnout. Physical features of burnout can include weight loss, bowel disorders, and higher resting heart rates.

What is clear beyond all is that the symptoms of burnout are very real for those who suffer their effects. Indeed, the repercussions can be serious for both the social and emotional health of the young athlete (as well as adult sportsmen for whom burnout is equally an issue). The basis of this psychological fatigue has been suggested to emerge as a result of four key elements[25] – an overwhelming of one's personal resources, negative self-appraisal, physiological stress responses, and inadequate support/coping strategies (see Figure 6.1).

An overwhelming of one's personal resources essentially refers to the question of whether one has the physical and mental capabilities to cope with a particular challenge? Clearly, if an experienced junior athlete is asked to compete at a level below his or her own there will be ample physical and mental resources

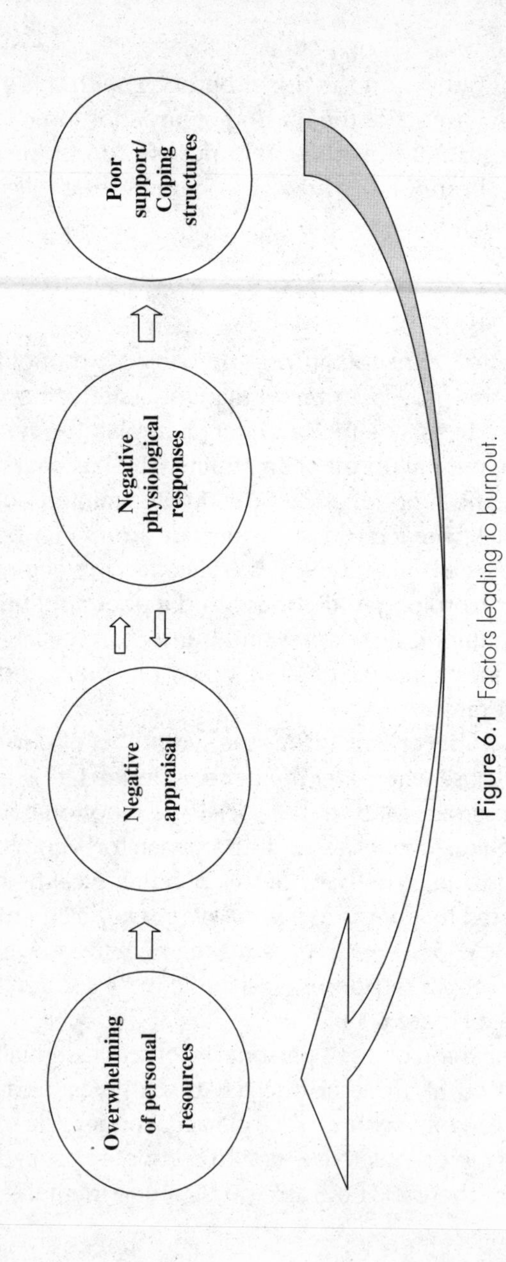

Figure 6.1 Factors leading to burnout.

available to deal with the challenges. However, if the same athlete is thrust into the age group above, this degree of resource abundance might not be so apparent. The subsequent stress experienced by the young athlete is mediated by this relationship of personal resources and upcoming challenges. It is worth stating that stress can also form when personal resources greatly outstrip the challenges. After all, who can stand boredom for very long without becoming frustrated?

In either case this is not the final word with regard to burnout. Many athletes are of course capable of keeping their spirits high despite facing insurmountable challenges much of the time. Accordingly, the second factor contributing to burnout acknowledges that individuals will often appraise the kinds of stress they face in a unique way. Indeed, two young swimmers who are entered into a higher age group for local competition might both suffer several bad defeats, at least relative to their successes at the younger age group. One might come to the conclusion, however irrationally, that his abilities are being called into question, *'I'm too rubbish to compete with these guys!'*. However, the other (more rational) youngster may well recognise the uphill battle he has been facing and accept that the circumstances – not his ability – are the root cause of his defeats. That is not to say one can endure loss after loss and explain it away in such a manner. Merely that each of us has a unique perspective in which we understand the events of our lives and what they signify.

The upshot of being placed in a situation that overwhelms one's personal resources, coupled with a tendency towards negative appraisal, can lead to adverse physiological responses. If one senses a threat – for a young athlete, a series of humiliating losses most definitely constitutes a threat – one is likely to become withdrawn, agitated, and angry. These are prototypical responses to danger (in this case a social danger – losing one's place in the pecking order) and are designed to aid the individual in such scenarios. Think of the fight or flight response discussed in the previous chapter; the loss of a match in this example might provoke a 'rage' that prompts one to perform better next time. Yet, while this makes sound evolutionary sense, if it endures it can a have a debilitating effect for the young athlete. After all, while brief moments of anger can work wonders for one's energy levels, who wants to feel negatively energised for an extended period? Rather than remedying the problems it may make them worse. One might go on to make even more negative appraisals of one's performances, as well as rejecting the help of those around – parents, coaches, and friends.

The clearest remedy following the observation that a young athlete is suffering the symptoms of staleness from his or her sport is time out; perhaps the chance to explore new sporting pursuits for a period of time or even just a complete break from physical exertions altogether. For a young footballing talent who has become disaffected with team practice sessions the switch to basketball or gymnastics for a brief respite can keep the levels of fitness up, as well as provide the opportunity to explore new pastimes and meet with friends. Ronnie O'Sullivan, world snooker champion, has on occasion expressed his need for such a respite – *'[snooker's] not making me happy, mentally and physically it's taking its toll… I probably won't play next year. I'll take a year off'*.[26]

But while the emergency remedy of enforced time away to treat the signs of burnout might well be necessary in extreme cases, this obviously is not a desirable outcome. The job the coach is paid to do is to maximise the performance of the player. Time away from training is not going to accomplish this goal, and an athlete suffering burnout can be indicative of a poor coaching system. Fortunately, for the young hopefuls who dream of sporting stardom, sport science has wised up to this dilemma.

A number of approaches now exist that aid in the enhancement of performance while guarding against burnout.

Let children play

In decades gone by, young athletes were often treated as mini-adults. They would have been expected to play football matches on full-size pitches and use tennis rackets designed for adults (perhaps with a cut-down handle, but nonetheless still very heavy). Competition from an early age has also been a feature of child sports in recent decades. And the drive to win medals and obtain ranking points has lead to a range of inappropriate coaching methods. Modern science has shown this to be a misunderstanding of both the physical and psychological changes that occur across childhood and adolescence. Children are neither as emotionally nor as physically capable as adults. Their minds and bodies are still going through great changes and accordingly it is necessary to adapt training methods to these needs.

Among the recent changes in youth sports training has been the introduction of programmes such as Istvan Balyi's Long-Term Athlete Development plan (LTAD),[27] which takes into account the changing psychological and physiological requirements of young athletes. And while this programme has been designed for maximising athletes' general sporting performance, the LTAD also illuminates valuable lessons with regard to avoiding burnout and maximising the psychological well-being of young athletes.

The LTAD plan typically advocates five stages of training, each tailored to the growing needs of the child – (1) the fundamental stage, (2) the training to train stage, (3) the training to compete stage, (4) the training to win stage, and (5) the retainment/retirement stage. Of these stages, the first four are of particular interest here, for they concern the early years of development up until the age an athlete will turn professional.

The fundamental stage typically involves children between the ages of 6 and 10. Here, the emphasis is on enjoyment of the sport alongside the teaching of basic coordination, agility, balance, and speed. Competition is not yet considered to be an important factor. The training to train stage is centred upon children between the ages of 8 and 12 (girls will commonly reach this stage slightly earlier than boys) and it is now that the sport-specific introduction begins. Young athletes are now taught the fundamentals of their sport – tactics, techniques, and general training methods – upon the platform provided by the fundamental stage. In the training to train stage, both competition and winning are encouraged, but the emphasis stills lies with the learning of the core skills and not competitive results. The training to compete stage builds upon both of the previous stages and involves young athletes between the ages of 13 and 18 (again, girls will mature slightly earlier than boys). Here, technical and tactical skill development is still a prominent objective of the training agenda, but specific competition knowledge is strongly incorporated into the programme.

The impact of the LTAD on preventing sports burnout, especially in the pre-teens, is perhaps obvious, particularly when one recalls the key factors underpinning burnout. The first factor – an overwhelming of personal resources – is largely eradicated under the LTAD as early competition is mostly sidelined as a non-essential feature of the young athlete's development. Instead, the focus centres upon progressing the child's skill set, which in many cases can be achieved without the need to inundate or undermine the sense of competence the child possesses. This is a view shared by the Spartak Tennis Academy in Moscow, which has produced a string of top ladies into the world game, '*none of [the] students are permitted to play in a tournament for the first three years of study'*.[28] This might seem overly cautious but the results from Spartak are hard to argue with. A notorious hothouse, the Nick Bollettieri tennis academy, has also adapted its teaching methods to take into account the psychological needs of young athletes, *'looking back to those early days it's easy to see how we may have sacrificed fun for competition. But the whole atmosphere at the academy has undergone big changes since then'*.[29]

Periodisation

The LTAD plan is not the only advance in training scheduling that has been adopted in recent years by elite coaches. The advent of the Soviet Bloc's sporting successes from the late 60s opened many Western eyes to a number of new training methods. One such method – periodisation – was soon the worst-kept secret in the coaching world, as it became clear that the physical gains one could achieve far outstripped the traditional methods, and, in addition, this method guarded against training apathy.

Periodisation simply means breaking one's training schedule into distinct blocks, each specialised to tap into a particular aspect of the athlete's wider development.[30] A typical periodised training schedule would be likely to include (1) a preparation phase, followed by (2) a pre-competition phase, leading into (3) the competition phase, and concluding with (4) an active rest phase (see Figure 6.2).

The preparation phase would be likely to vary from sport to sport in actual content. However, in general this period would aim to develop base fitness and/or modify techniques without the need to face the pressure of competition with these new skills. To a young boxer this phase might represent the opportunity to learn to slip the opponents jab without the fears of having to take the unlearned technique into a competitive situation

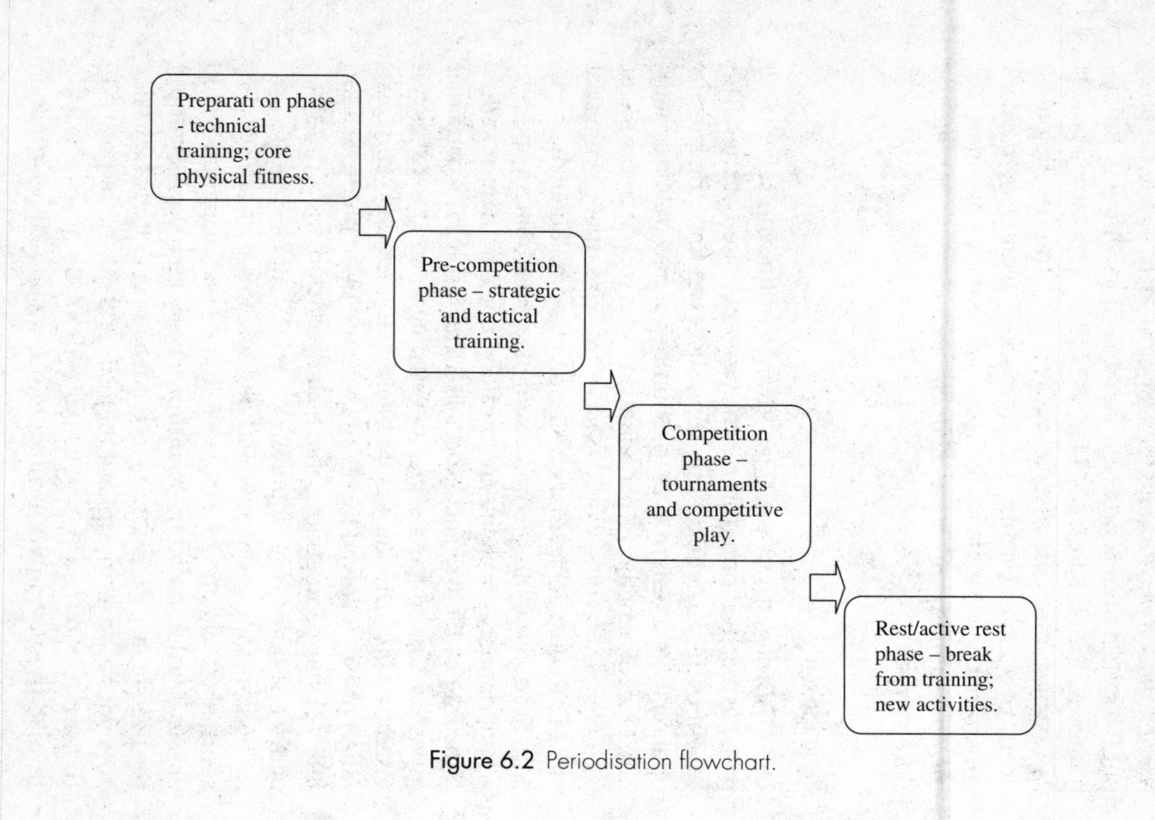

Figure 6.2 Periodisation flowchart.

where he might come unstuck and taste a painful defeat. For Ricky Hatton's trainer, Billy Graham, this part of the training is '*about working on…[and] developing his fitness, which is still important at this stage*'.[31]

Following this one would begin the pre-competition phase, where attention is turned towards tactics and strategies that are designed to prepare the athlete for competition. Here, the young boxer might learn how to implement his new technique of slipping the jab against a range of opponents. Can he take advantage of this new skill by exposing the chin of the opponent as they attempt an unsuccessful left jab? For Ricky Hatton this is the stage where he begins to combine his basic training with his boxing skills. '*The most important aspect of Hatton's training regime is introduced – the body-belt, a heavy, rounded, padded sack which Graham straps on and Hatton punches, designed to hone his fighting qualities: body-punching and sheer relentlessness.*'[32]

The competition phase normally involves only moderate maintenance work to keep fitness at a peak and plenty of specific competition preparation. '*All we need [now] is a ring, a body-belt and some sparring partners,*'[33] says Billy Graham of Hatton's pre-fight final preparations.

Upon completion of the competition phase the all-so-important transition/rest phase is typically scheduled for a time span of a few days to several weeks to allow the athlete a chance to reflect and take stock of what the successes and pitfalls of the preceding weeks and months of training might have been. At this point the goals and aims of the next set of training blocks will be set and the effectiveness of the one just passed will be assessed.

The benefits of a periodised training regimen are clear when one understands the basic details of how the body responds to physiological stress. Much of the knowledge stems from the work of Hans Selye,[34] who argued that the body has a three-stage response to stress (here, sports training stress). The first stage is *shock* – the familiar muscle soreness/mental fatigue one experiences following an unusually hard training session. The second stage is *adaptation* – when the body/mind adapts to the nature of the training stresses. The third stage is *staleness* – when adaptation no longer occurs, as the muscles/mind are no longer being stressed sufficiently to encourage growth.

With regard to Selye's insights, the benefits in developing a periodised training programme for elite young athletes are considerable. Indeed, in years/decades gone by the favoured regimen of many sports coaches may have been somewhat repetitive – '*get out there and give me twenty laps!*'. This 'good old-fashioned' approach is still the default option for many. However, it neglects the manner in which humans respond to training, and as such, periodised regimens allow hard physical and mental training to be conducted without pushing the young athlete into the third phase of Selye's model – staleness – a likely precursor to full-blown burnout.

The gradated approach of periodisation also allows one to develop skills and abilities in the specified stages without the consistent glare of competitive pressures. By careful scheduling the young athlete is not in danger of having his or her personal resources overwhelmed (the first factor leading to burnout). Furthermore, the mindless repetition in training regimes of decades gone by is now, at least partially, reduced (there is of course no way of removing the need for hard graft within the training programme all together). Accordingly, for young athletes who have limited attention spans and would have struggled to maintain interest in rigorous training of a more repetitive nature, the boredom factor is lessened. Indeed, even those who truly do love punching the heavy bag or running shuttle drills will reach their saturation point at some time. Periodised programmes offer some protection against this staleness.

Goal-setting

We have seen how the LTAD plan and periodisation can help to deliver the appropriate doses of training stress to a young athlete. In doing so, the over- and underwhelming of personal resources is reduced and the likelihood of burnout is lowered. However, the danger of negative appraisal still lingers. Accordingly, there is a need to minimise this destructive process emerging.

A sure way to encourage negative appraisals from a young athlete of his or her performances is to leave the goals unspecified; how can they know whether or not they have achieved if they didn't know what the target was to start with? To this end, goal-setting is perhaps the training technique that most prominently illuminates exactly what is required with regard to results and performance. However, the process of setting goals, particularly for elite athletes, is somewhat of an art form. Indeed, striding into the gym with the aim of 'improving one's 10k time' or 'achieving a better serve' are noble in intention but often next to worthless.

The reason for this is largely obvious, although often ignored; such goals do not specify a course of action that will help one achieve them, as well as being so vague that it is unlikely one would even know what the desired outcome was. Contrast this relative naivety with the approach of leading tennis coach Brad Gilbert – '*here's my 100% coaching principle… I don't talk about numbers; I don't obsess about rankings or results. What I talk about are those specific things we need to do to keep getting better*'.[35]

Gilbert implicitly refers to what sports psychologists call *process goals*.[36] These goals concern tangible and measurable aspects of an athlete's blueprint for improvement. They are the incremental changes one needs to make in order to become better. For a young sprint hurdler this might be straightening one's lead leg over the hurdle or holding one's form better on landing. Typically, a course of action will follow these process goals, such as specialised exercises that facilitate the improvement. They will also operate over a relatively small timeframe, so that they are manageable and do not become overwhelming.

While these process goals can help to inspire a young athlete to persevere through tough training sessions, they ultimately serve to feed into what are known as *performance goals* and *outcome goals*.[37] Performance goals are the next rung up on the goal-ladder hierarchy. For the young hurdler above, whose process goal was to improve her hurdling technique, a performance goal can be viewed as the culmination of this aim – such as bettering one's personal best by 0.5 seconds. Finally, outcome goals are the highest rung in the hierarchy. One's outcome goals can be seen as the joint culmination of achieving both the process goals and the performance goals, such as winning a national championship sprint hurdles medal.

Olympic champion sprinter Michael Johnson has spoken of his own goal-setting process, which incorporates process, performance, and outcome goals. *'At the beginning of the season, I sit down and talk with my coach and with other people whose opinions I value. And I think about what I want to do that year. When I've identified specific goals, I write them down on a piece of paper. Then I ask myself… "How are you going to do that?" That's when I connect the dots, stringing plans between the goals.'*[38]

Clearly, such a hierarchical structure is superior to challenges set on a whim. They promote meaningful action towards improvement and allow an athlete to measure their progress and feel satisfied that they have achieved something worthwhile. This in itself reveals another important aspect of setting goals for young aspiring athletes; inherently, they must be achievable. While it is commendable to expect a promising local swimmer to win the national championships, if she has yet to win her county title what hope has she of making such a radical jump? And while she may make the grade, more often than not failure will hang over her head. Referring back to the causes of burnout, such overambition might overload the youngster's personal resources and prompt her to negatively self-appraise, two out of three components in the model of burnout above. Nonetheless once the formalities of setting meaningful and achievable goals are established, the likelihood of this occurring is lessened.

Lean on me

Once the finer points of a training programme have been identified and implemented, the danger of burnout still exists. Recall, if you will, the model of burnout described above. The trials and tribulations that can arise from pitting one young competitor (or team) against another are serious psychological challenges for developing minds, despite well-structured training regimens. The need for positive support structures – family, friends, and coaches – is, therefore, an essential component in creating a positive environment for a young athlete to learn and progress within his or her sport.

However, this aspect of a young athlete's life can often be less than favourable. All who have had some involvement with youth sports will know of those parents who, despite perhaps better intentions, hinder their child's progress through their overzealous berating of officials, and even other junior competitors. And there is of course the infamous parent who purportedly lives through the successes of the child as a means to experience elite success that perhaps was unattainable in their own childhood. This occurrence has become so well noted that it has even acquired its own name – the *reverse-dependency trap*[39] – signifying those parents who have become dependent on the success of their child, even defining their self-worth upon their offspring's victories.

The dangers of such close involvement in the success of a child are perhaps self-apparent. When the approval and affections of a parent are somewhat conditional on sporting results it is likely that the relationship will be stressful. In turn, this may lead to a breakdown of support in crucial moments. The former pound-for-pound king of the boxing world Roy Jones Junior has spoken on this subject with regard to the volatile relationship he has endured with his father, '*having him [in the corner] was good, but it was more bad than good… It caused too much havoc*'.[40]

The challenge for the coach in such moments is to attempt to communicate the dangers of overinvolvement, or in the case of the absent parent, to illuminate the benefits that emotional and practical support can have in maximising the potential of the child. Karolj Seles is one who is credited in providing the right forms of support to daughter Monica. Over and above his ability to make his daughter laugh at his tennis balls adorned with her favourite Disney cartoons,[41] Monica has commented on his tremendous emotional support. '*He saw the bigger picture of sports, instead of just win or lose. He was human. Sports is a business and cutthroat and people will do anything to win, but I was lucky I had my dad as my coach.*'[42]

The same sentiments have been spoken of former England and Barcelona football manager Bobby Robson – '*Everyone loves [him]. He was an infectious personality, and that made you want to play for him. When he was England manager he was always fiercely loyal to his players, and was never swayed by the media or the fans. He knew who his best players were and he stood by them even when things didn't go right.*'[43] Similar words of praise have been heaped upon Manchester United's manager, Alex Ferguson for his powers of nurture – '*He just gave players so much belief and even when we played Real Madrid in that Cup-Winners' Cup final he wasn't fazed at all and made sure we weren't either. His enormous mental strength is unquestionable.'*[44]

Conclusion

The professional world of sport demands an enormous chunk of an athlete's personal resources. One might be away from family for long periods, be required to go through the pain barrier on a daily basis in training, and have one's abilities placed under scrutiny to an all-too-frequently unsympathetic public. The results of such pressures can be the burnout syndrome and nowhere is this more apparent than in youth sports. Children need to train prodigiously hard if they want to make it in the ruthless world of professional sport. Yet they often lack the emotional maturity to endure the rigours of hard training.

While the hard graft and sacrifices cannot be removed from the training process, there are tools at the coach's disposal for minimising the likelihood of burnout. As we have seen, taking into account the age and maturity of a young athlete and adopting an appropriate training plan can reduce the pressure of overwhelming a youngster who is still learning the basics of his trade. This, coupled with the tool of periodisation, can significantly reduce staleness and subsequent burnout. The use of focused and specific goal-setting practices can limit the negative appraisals that frequently follow a poorly perceived performance. Goals allow one to see the 'wood for the trees'. In addition, the role of a nurturing support team – a young athlete's parents, coach, and friends – is invaluable in allowing moments of vulnerability to arise without long-lasting dips in confidence or motivation. Indeed, if a young athlete knows they can attempt to learn a new and difficult technique without incurring criticism as they take the inevitable step back in order to take two forward, they will be considerably more likely to succeed in their attempts.

The previous chapters, this one not withstanding, have afforded an insight into many of the crucial influences that underpin the making of an elite sporting mind. But what of the true life stories of elite athletes? Does the science truly match the reality? Chapter 7, 'The Lives of Others', asks just this question and explores the lives of some of the world's greatest athletes in search of an answer.

7
the lives of others

Quite some ground has been covered from the first tentative questions posed in the first chapter regarding the origins of elite sporting psychology. More than anything, I hope that a sense of the tremendous complexity involved in the development of sophisticated abilities in an organism as complex as a human being is now only too apparent. And of course the elite sporting brain is no exception.

Yet many of the claims reported thus far, despite largely being grounded in empirical observations, are open to criticism for being somewhat detached from the real lives of those elite athletes whose skills are under the microscope. After all, research conducted in the lab simply isn't the same as the complex realities of the outside world, with all of its complications and inconsistencies. Scientific theories are appealing for their ability to explain confusing phenomena. However, instinct may implore one to challenge the ideas that are banded around, and those in the preceding chapters are not immune to this kind of challenge.

This form of criticism is of course not new to those who explore psychological phenomena. Indeed, the concept of ecological validity – the sense of gauging just how realistic one's findings are to the sensibilities of the real world – is taken seriously by the scientific community. And so, it falls to me at this moment to attempt to ground the arguments I have proffered thus far in evidence more appealing to the sceptic than the reports of scientific experiments alone.

As both quotes allude to at the head of this chapter, talk is cheap – actions speak louder than words (although one should never aim to replace scientific rigour with anecdotal evidence, rather one should use it as a complement). Accordingly, in the sections that follow an attempt has been made to ground the arguments of the previous chapters in the 'lives of others'; namely, the stories of some of the world's elite athletes. In short, the challenge I set myself was to see whether or not the life stories of champions, such as Tiger Woods, Ronaldinho, and Andre Agassi, among others, really supported the theories that claim to explain elite developmental trajectories described in the previous chapters.

Into the looking glass

Before we delve into the early lives of elite athletes looking for corroborating evidence to the theories, let us quickly remind ourselves of the details one would expect to have occurred throughout the development of a champion's psychology. The preceding chapters have suggested that a host of possible influences, set against the backdrop of a highly interdependent nature–nurture relationship, are crucial in forming an elite sporting mind. It is clear that no one feature will stand out above all others. Rather, a cluster of favourable inputs to the development of the elite athlete is what one would expect to be present.

One would expect the early years of a future champion to have shown the basic tendencies of what would later become the elite sporting personality traits; extraordinary tenacity, an unusual attention to detail, high drive, among certain other characteristics. One might also anticipate a family environment where the dynamics were conducive to nurturing and harnessing any innate giftedness. This might arise through the apparent favour of being firstborn (under the view that such children are nurtured more diligently), or for more complex reasons that conspire to place the child in a developmental climate that encourages and supports sporting success. One would also be likely to observe the elite athletes' early lives containing exposure to an elite peer group, one where an elite attitude to training, alongside the opportunity to develop one's skills against strong opposition, was available. And with this elite environment, the nature of the future champion's coaching will of course impact upon his or her development; the Pygmalion effect – where the expectations of a teacher can be a self-fulfilling prophecy for the pupil's later successes – is testament to this influence.

In addition to these more formalised influences, elite athletes may well have experienced anomalous influences that provided an additional 'nudge' towards greatness. Perhaps something as simple an early opportunity to watch a hero in action in the flesh may have presented itself, or the good fortune to be selected for a prestigious training squad against the odds, impressed itself upon the young aspiring athlete.

There is certainly no prescriptive formula for elite success, an important detail highlighted previously. However, a broad base of fundamental influences can set off the hopeful champion in the right direction, even if the path is still sure to be fraught with uncertainty. With this thought in mind, let us take a journey through the early years of some of the greatest athletes the world has ever known.

Basic tendencies

As we now know, a key early shaper of an elite sporting personality is to be found in one's basic tendencies. A number of striking observations emerge when one explores this aspect of elite athletes' early lives. The Romanian gymnast Nadia Comaneci, winner of five Olympic gold medals and the first gymnast in history to score a perfect 'ten' in Olympic competition, has said, '*if ever a child aged its parents overnight, it was me*',[1] on account of her insatiable energy and curiosity. Indeed, the young and persevering Nadia managed to bring down the family Christmas tree at the tender age of two in a relentless quest for a tempting sweet dangling near the top. When her father rescued her from underneath the toppled branches he found her still clutching the sweet somewhat overzealously.

This anecdote of young Nadia's relentlessness and single-mindedness is perhaps somewhat more striking alongside her account of being struck by a female gymnastics coach following a fall at the national gymnastics championships when she was just nine. This would be a disturbing moment for most young athletes. Comaneci's own appraisal of the incident was that she *deserved* to be hit, such was her own disgust with herself at the time.[2] Comaneci seems to have been so unusually driven and extremely critical of her performances to the extent that this appraisal is (maybe) not so surprising.

Comaneci has also spoken of her early 'aloofness' – not mixing much with others – but has likened this quality to a concentration in things that were important to her, rather than as anything more worrying. At the age of six, she reportedly scared her parents half to death when she failed to return from a solo outing until late in the evening, having become distracted with her own thoughts and wandered through the fields near her home in a reverie.[3] What is perhaps more intriguing than the ambivalence towards the concerns of others (after all, how many young children think very far ahead in time) is the apparent independence exhibited at such an early age. Comaneci, as judged by her later success, was nothing less than a phenomenon, standing apart even among the elite of the sporting world. And the early signs of such success, a total single-mindedness, independence, and high focus, emanate from her childhood years.

Tennis superstar Steffi Graf's early character has frequently been spoken of in a vein similar to that of Nadia Comaneci. The young Steffi was intrigued with the sport of tennis – she saw her father Peter playing, who was an avid enthusiast – and was so persistent that she should be allowed to play that Peter eventually tied a string over two chairs at home so he could satisfy her demands.[4]

And her enthusiasm for the sport showed no sign of dampening following this early foray into the game. Peter was soon finding himself out of pocket as he challenged his precocious young daughter to achieve fifty tap-ups with racket and ball so as to win an expensive German pudding. Steffi persevered and won the prize. She also conquered many of the further challenges set by her father with an unusual concentration. She was said to be indifferent to the ringing of the phone or doorbell if she was deep into a rally with her father, even as a very young child.

Her enthusiasm and dedication in her love of tennis are strikingly illuminated by her mother when she commented, '*I had to force [Steffi] to do things which other girls of her age usually do, like play with dolls and go to friends' birthday parties*'.[5] This sentiment is echoed in the views of the president of the local tennis authority at the time of Graf's first forays into the tennis world. '*She was very serious… and never had to be forced to play… she would only cry if she had to finish early.*'[6]

Cricket legend Ian Botham has also spoken of his astonishingly precocious early years. From his very first days Botham has commented, '*I had an unshakeable self-confidence and belief in my own abilities.*'[7] He has also spoken of his relentless curiosity of novel things, which in turn sparked many conflicts with his parents.

One such conflict was Botham's incessant wish to cross the road near his home alone while he was still very young. Indeed, this was something he recklessly attempted in a momentary lapse of concentration by his mother and was spanked for his efforts. Botham also reportedly barged straight into fellow competitors at school sports day races in an attempt to scupper their chances, such was his competitiveness and desire to finish first. Somewhat humorously, these tendencies prompted the local doctor to suggest that the young Ian wear protective headgear in his daily play following a hospital trip for a bang to the head.

And it appears that similar stories emerge from a large percentage of future elites. Golfing superstar Tiger Woods reputedly memorised his father Earl's work phone number at the age of two so he could call him at work and ask him to play golf with him, an act that became a routine between father and son as the years rolled on.

Pound-for-pound boxing great Roberto Duran was said to have been so enamoured with the sport of boxing that when he first learned the noble art as a pre-teen he would sometimes skip sleeping so he could train on the beaches near his home in Panama, despite the fact that his family had little money with which to buy food.[8] This of course was before any real notion of a future in boxing could be appreciated and powerfully illustrates the type of motivation Duran had towards his beloved sport. (It is worth mentioning that Duran was in many ways a street kid, growing up in a poverty-stricken neighbourhood in Panama, and so a night out under the stars was perhaps not the extraordinary occurrence that one might expect in the 'First World').

While many boisterous young boys might share Ian Botham's reckless approach to life, it is a child somewhat more prodigious who displays the kinds of tenacity, drive, and single-mindedness that eventually contribute to sporting glory at the highest plateaus. Very few will be able to recall single-mindedness similar to that of Nadia Comaneci, or an enthusiasm like that of a young Roberto Duran, in their own early childhood. And so, it appears that the extraordinary sporting feats of later years were in many cases preceded by extraordinary basic tendencies that laid the early foundations for an elite sporting trajectory.

Family affairs

These early basic tendencies are of course little more than a footnote in the life of a sports-mad child if other key sculpting factors do not accompany them. The role of the family has long been argued to provide a pivotal developmental influence. In particular, birth-order effects are another factor claimed to shape an individual's development. However, as we have seen previously, this appears to be a pseudo-explanation. The reasons for this effect seem to emerge from factors such as an increased exposure to cognitively demanding environments (a wholly adult environment is what a firstborn experiences – at least before being joined by any siblings) and the greater investment parents might make in their firstborns. As was noted in chapter 3, these favourable factors are not necessarily exclusive to firstborns. The influence that seems to be most fundamental within the dynamics of the family appear to be less about the birth-order and more about the expectations placed upon the child.

The greatest golfer of his (and perhaps any) generation Tiger Woods is living testament to this notion. Tiger's father, Earl, had already fathered three other children from a previous marriage and had happily settled into a life post fatherhood; he had no wish to raise another family. However, Earl found himself falling in love with Tiger's mother, Kultida, and soon embraced his chance to become a father again. Indeed, Tiger appears to have become a talisman for the middle-aged Earl, a second-chance at fatherhood, *'the fact I had three children was like [God] saying, "I'll give a trial run. Let him have some children and see how he handles it, but he's got to be able to do everything. I want him to know how difficult it is because I want the best for Tiger"'.*[9]

Box 7.1 An insight into an elite sporting mind

Andy Nicol – Rugby Union – former captain of Scotland and Bath

Basic tendencies/elite personality:

'My mum would tell you I was a nightmare baby... she put it down to [being] hyperactive... [I was] notoriously falling out of trees...'

'There are huge sacrifices you have to make to play a sport to a high level... I can't remember what it was that made me so committed, but it just felt totally natural... I worked incredibly hard, I put so much into it'.

'I don't know why I was so mentally tough, but I always was... I was the player others looked to for inspiration... as a leader'.

Exposure to elite training environment:

'I started playing mini-rugby when I was six... it was very much in the family... my grandpa played for Scotland... my older brother played for Scotland U/18's...it was very much expected that I would be playing a lot of rugby'.

'I swam a lot and played a lot of football [on top of my rugby]... very competitively until around the age of fifteen... I got to the national finals [as a swimmer]... it was very much an elite environment... there were a number of Commonwealth Games medallists... I was training before and after school... I had a very good aerobic base when playing rugby in later years... you wonder whether the swimming gave me this quality early on'.

'I was lucky... I had a talent but also the environment... I think the swimming started my drive and taught me how to train like an athlete'.

'I got asked to train with Dundee football club as a youngster... football gave me a real sense of coordination and balance... in rugby you often just put your head down and run... in football you have to look up... it really taught me to look up... in my position scrum-half this is a crucial skill... football really helped me with this'.

Support structures:

'There was an incredible commitment from my parents getting me to and from the many training sessions'.

'I had one coach, the school P.E. master, who was an enormous influence on me... he really took me under his wing and made me the player I ended up being... he would do sessions with me at lunchtime... we would pass the ball in the gym... a huge influence on me'.

Mental Toughness:

'That five minute period before a big game, I just hated it... because that's when you question yourself, your fitness, the ability of your team to perform... it was the last place on Earth I wanted to be. But five minutes later there was no other place on Earth I'd rather be... they say nerves are good because they focus the mind, but anxiety is bad because it dulls the senses'.

'I was always conscious about being in the right frame of mind before a game... that said, it always came naturally to me... but knowledge is absolutely everything... I knew what I was and wasn't good at... all my career I've prided myself on preparation and being in the right frame of mind'.

This extraordinary reverence held by Earl for the young Tiger seems to have made certain that he was going to be nurtured in a positive direction (at least by intention – the methods undertaken by Earl are discussed shortly). The fact that golf emerged as the medium of choice was perhaps an accident, Earl falling in love with the game late in life, shortly before Tiger was conceived – '*I decided if I ever had another child, he or she would be exposed to the game earlier than I had been. And along came Tiger.*'[10]

Earl Woods, for reasons particular to him, seems to have been overcome with an enormous enthusiasm for the birth of Tiger (over and above that which one might ordinarily experience). And a similar story emerges in the case of cricketer Ian Botham. Prior to his conception Botham's parents had tried for some time to conceive of a child and had had the misfortune of suffering a number of miscarriages. His healthy birth thus signified not only the first child of his parents, but also a child who was dearly longed for following the medical difficulties his mother had faced during her previous pregnancies. Botham seems to have (perhaps unsurprisingly) been bestowed with a special kind of significance to his mother and father as a result of their trials and tribulations in becoming parents. '*Probably no child was ever more eagerly anticipated nor more dearly loved. Their baby could do no wrong and throughout my childhood I was indulged shamefully, spoilt rotten, in fact.*'[11]

Such stories of course neglect the fact that many more typical family rearing episodes are less dramatic; nonetheless, the drive and expectations of the parents for their children still appears to be a ubiquitous presence in the formative years of the elite athletes. John McEnroe, Wimbledon champion, was raised in a family who, although frequently enjoyed a raucous party befitting their Irish roots, simply didn't tolerate anything less than the finest work ethic from their brood – '*my parents had a serious and demanding side. They expected achievement*'.[12]

McEnroe has illustrated this 'tough love' approach with an incident of his childhood following a tumble off his bicycle. His mother, who was a nurse, tended to what she thought were just bruises and, not wishing to indulge the young John in sympathy, sent him off to the tennis courts to get on with his practice. The fracture, and not mere bruises, he had received only transpired three weeks later, *'My mom and dad were strivers in every way; they fully bought into the American dream.'*[13]

Michael Johnson, Olympic gold medallist at 200 m and 400 m, has spoken of a similar family background. Raised in a family proud of their hardworking ethics, Johnson's father was a perennial stickler for challenging his children to do their daily tasks, however menial, as best they could. And he encouraged them to come to him when they had an accomplishment they wished to achieve with a detailed plan of action. This moral seems to have rubbed off on the Johnson brood. Michael, somewhat humorously, tells of the time he was close to failing a math course in college, and how he decided, in line with the kind of planning his father endorsed, to go to bed at 8 pm so he could study early in the morning when he was mentally most fresh.[14]

Standing on the shoulders of giants

From the evidence of the previous chapters it is only too clear that a fortuitous predisposition to a sporting personality, and a family that expects excellence and strives to promote it with their children, are important but not sufficient factors in producing an elite sporting mind. The exposure to an elite environment, one where a promising young athlete can measure themselves against other promising rivals, develop advanced tactics and perceptual skills, and hone a steely competitive instinct, is another key influence that needs to be present to sculpt an elite sporting psychology.

Steffi Graf was one such budding prospect that seems to have totally epitomised the notion of right place, right time. Steffi's father, Peter, who had nurtured her early start in the game, recognised his daughter needed more advanced training (at the age of 5) and began a hunt for the most suitable coach. The choice of Peter Graf was the Baden Tennis club in the town of Leimen, under the direction of the former Yugoslavian Davis Cup team member Boris Breskvar, which was a hub of elite junior activity at the time of Steffi's early development. The testament to the potency of the Baden academy is illustrated in the observation that Steffi's male counterpart Boris Becker, was also a product of the Breskvar training regimen. Becker, of course, occupied the throne of men's tennis following his Wimbledon triumphs that catapulted German tennis into the world's sporting consciousness.

The emergence of the two greatest players in German tennis history from the same club at the same moment in time is possibly sheer coincidence. However, the sophistication of the training these young budding stars were exposed to from early on in their playing careers suggests otherwise. Breskvar collaborated with Hermann Reider, professor of sport science at Heidelberg University, and together they implemented a regime that targeted not just the techniques and tactics of tennis, but the development of psychological and physical abilities. '*For five years he helped me with Boris and Steffi, making psychological tests, motivational tests,*'[15] says Breskvar of Reider.

The coaching team also incorporated many other ball sports, believing that developing a sense of the dynamics of play was a key facet of sporting expertise – '*this is very important when children are 9, 10, 11, because you must play a lot of combinations … if you can transfer this to tennis, you can improve a lot. Steffi is a wonderful basketball player. Boris is good in basketball and very, very strong in soccer*'.[16] The sentiment that stands out here is that the young talents in this academy were not only receiving excellent daily tuition in their stroke play. They were also receiving elite perceptual and decision-making skills training by virtue of the inclusion of other ball sports that tapped into the requirement to recognise patterns in play and act accordingly. And while this approach is not perhaps unusual for young athletes, it seems that Breskvar was more conscious than most in using these other sports as platforms to developing tennis skills.

Steffi's exposure to elite training was further enhanced while she was still not much taller than the tennis net when her father, Peter, moved the family to a sleepy town that happened to have three dusty tennis courts in the local vicinity. Peter rented these courts out and supplemented his income by giving tennis lessons to the local children. The greatest recipient of this tuition appears to have been his daughter, Steffi. Indeed, Peter was perhaps as driven as Steffi herself, and resolute in enhancing her game. Peter took notes of her lessons and incorporated what he saw so as to make every session he spent with Steffi novel and designed to improve a particular facet of her game. Peter's keen eye for the details extended to scouting other players with his daughter, both on the television and at junior tournaments, where they would analyse the strengths and weaknesses of potential rivals of both present and future. There can be no question that Steffi's formative years were filled with training designed and implemented to bring the best out of a young tennis hopeful.

Tiger Woods was another future champion, who was exposed to the kind of training environment that fosters the rise to greatness. Earl Woods's ambition to provide the very best for the son he felt was a divine blessing, coupled with his own enthusiasm for the game of golf and his wish to share this enthusiasm with his son, has been described above. However, this parental desire does not often transpire into the child reaching the professional game, let alone the lofty heights that Tiger Woods has sustained in the golf world. One of the key differences in this tale is the story of Earl's own background and of how it conspired to provide an outstanding platform upon which the young Tiger was able to achieve his goals.

One important observation that has already been made above is that Earl fathered Tiger late, when he had already experienced the joys and rigours of parenthood thrice before. Thus, he knew what to expect, and in many ways was no doubt able to 'hit the ground running', so to speak. But what perhaps is equally as crucial a factor in Tiger's early development is the career and experiences that Earl enjoyed outside his parenting responsibilities.

Earl has spoken of his life-long teaching career, both as a little league baseball coach and also of his subsequent responsibilities in the army training young cadets, as an important prelude to training his son Tiger. The result of these experiences appears to have been profound, for whereas good intentions are one thing, the ability to see such intentions through is a considerable skill that many will never fully learn. And so the upshot of Earl's successful teaching career seems to have uniquely prepared him to nurturing Tiger in his crucial formative years. Indeed, an insight into the methods used by Earl Woods to sculpt the progress of the young Tiger is a fascinating tour de force in how to distil early promise into the finished article. One intriguing set of methods employed by Earl was in maintaining the motivation of Tiger, a fundamental component of any elite athlete's arsenal. Earl's advice is to '*always keep them wanting more*',[17] with regard to how much they play.

What this amounts to in practice is always agreeing to the child's desire to partake in an activity, but not allowing this activity to reach saturation for the risk of boredom setting in. This can be achieved by simply taking the clubs away from the young golfer after the fiftieth drive of the day, or by more ingenious means. '*[Tiger] was two years old... he would call me at the office... and ask, "Daddy, can I practice with you today?"*

After a pause where I had him worrying that I would say no, I'd say, "All right."'[18]

Similar methods were utilised in ensuring the young Tiger was not overwhelmed by the frustrations of learning a challenging sport, as golf undoubtedly is. Earl has advised parents of young athletes to, '*Avoid criticism. Never make negative comments. Positive reinforcement is much more effective... "I think you would have hit the ball much better if you played the ball a little more toward your front foot"*'.[19] This is a powerful, yet largely underutilised message. Indeed, how many parents a coaches routinely respond to a bad performance from a young athlete with a negative rebuke? This is not to say that poor behaviour or the like should be brushed aside, but merely that confidence and motivation are more likely to emerge from a teaching environment that fosters positive self-concepts.

The structure of Tiger's early practice sessions also owes a debt of gratitude to Earl's previous career. As Earl has noted, '*there is practice and good practice. Every practice session must have a purpose. You just don't go out to the course ... and blindly bang balls. There should be an order to the practice session.*'[20] The upshot of such sentiments resulted in the young Tiger being challenged by his father to tell him what his target was for every shot he played, '*each shot you hit in your entire life must have a target, or it is a wasted effort. Never hit a shot without a target. Why? The human body... must always have something to focus on, the more specific the better*'.[21]

If the extent of Earl's quality of teaching was not already demonstrable, his approach to instilling mental toughness in Tiger surely completes the case. Earl made the wise move (as one shall see) of establishing a code word so that if Tiger wanted the training session to end all he had to do was utter it, but that aside he was allowed to say nothing else during practice. With this assured, Earl took Tiger onto the course (around the age of 12) and proceeded to give him perhaps the most tempestuous training one could imagine. '*I pulled every nasty, dirty, rambunctious, obnoxious trick... week after week. I dropped a bag of clubs at impact of his swing. I imitated a crow's voice when he was stroking a putt... I played with his mind.*'[22]

Unsurprisingly, Tiger was frequently furious with his father. He even had to endure his cheating, and yet could say nothing about it by virtue of the deal he had agreed to before this mental training programme. And Earl has pointed to the experiences of Tiger's mental toughness 'finishing school' as a fundamental prelude to his tremendous match play record where one-on-one competition can prompt the jitters from even the hardiest of golf professionals.

Andre Agassi was yet another great champion whose achievements have been preceded by an extraordinary childhood environment almost solely devoted to the production of a world-class athlete. Mike Agassi, Andre's father, was an undeniable addict to the game of tennis, having first been exposed to the game in his native Iran by American G.I.s who he saw playing at the local army courts. This all-consuming love for the game led to a house viewing where his wife inspected the interior of the prospective property, while all Mike was interested in was the size of the garden. He wanted to know whether or not it was large enough to house a tennis court to support his obsession with the game; it was, and he duly built it.

The passion for the game also extended to his ambition that his kids should be champion players; he is said to have hung a tennis ball over Andre's crib and placed a cut down tennis racket in his hand when he was just a week old, presumably to attune him to the sport as early as possible. The methods became more extensive as time went on. The garden court was soon littered with tennis equipment. Numerous ball machines were employed to provide any type of spin or power shot one might need to face in a match. And they were used to endorse Mike's motto that 4000 balls in two hours is better than 2000 balls in four hours. The young Agassi clan were essentially raised on the court Mike built for this very purpose.

However, despite what might appear on the surface to be an excellent environment to nurture a budding tennis star, all was not well in the Agassi household. Those who recall Agassi's early professional years will remember that he was as famous for his flouting of convention as he was for his tennis, with his punk haircuts and ripped clothes. He soon became Nike's new poster boy, endorsing a unique brand of clothing decked with neon colours and Lycra spandex. And it was this rebelliousness that was also leading him to increasing showdowns with his father. The totalitarian regime he lived under became a serious bone of contention as he approached his teenage years.

Fortunately for Andre he was spotted by Nick Bollettieri, founder of one of the best-known academies in tennis, and offered a full scholarship. As such, Andre went to live in Florida and escaped the overbearing relationship he had suffered with his dad back at home in Las Vegas. However, the Bollettieri academy was perhaps tougher on their pupils than even Mike Agassi was known to be, with dawn training sessions that could continue to dusk with regularity. Yet, while this might have spelt disaster for a young prodigy who was already rebelling against the system on one side of the United States – '*he was a punk, a brat, a maniac, or whatever you want to call him*'[23] – Bollettieri showed uncharacteristic lenience with the young Andre for the many misdemeanours he partook in. It appears that he was careful not to kill the attitude in Andre that clearly marked him out from all the other young prospects. '*Yes, he was on the cutting edge several times. No way did I ever want to kill his personality, though. If I had killed Andre's attitude, I'd have killed the champion inside him.*'[24]

Stand by me

A mix of innate personality traits, favourable family dynamics, and the exposure to elite training methods are a fantastic start for any young hopeful's quest to reach sporting stardom. But beyond these facets, the emotional support and nurture of the coach and family is vital in allowing the difficulties that are inherent in a fledgling professional sporting career to be successfully negotiated.

We have already seen the considerable impact made by Earl Woods on the sporting development of his son Tiger, and this observation is not different when one notes the support provided by the young Tiger's family, as he made the difficult transition from junior prodigy to professional athlete. Tiger himself has noted that '*my family have been behind me from the beginning... they are my foundation*'.[25] Earl has been quoted in support of such sentiments, '*my son was subconsciously secure, knowing that whatever parameters we established, he could always be confident that behind him was parental strength and power*'.[26]

Ronaldinho, Brazil's World Cup winning footballer, was another athlete who was fortunate enough to receive his family's total support during the early years of his career. This was in many ways a peculiar twist of fate. Ronaldinho's older brother, Roberto Assis, was by some reports the greater of the two talents (an astonishing thought given Ronaldinho's standing in world football). However, the death of João, the father of the brothers, led to Ronaldinho's older brother making the career move of acquiring a transfer to the lowly regarded Swiss professional league so as to support his family. And while this move led to financial security for the family – a European wage in Brazil is considered to be a fortune – it marked the death knell for Assis's own future career aspirations at a big club.

This turn of events seems to have made Ronaldinho's mother, Dona Miguelina, that much more determined to give him the opportunity to reach his full potential, and perhaps in doing so also offer the family a way out of destitute poverty in the absence of the traditional breadwinner. '*Ronaldinho never had to put his clothes in the wash, tidy his room, help prepare or clear up after a meal... all these measures were designed to ensure [he] had no concerns other than his football.*'[27]

She also guarded against serious girlfriends, knowing that a relationship would deter her young son from accepting any potential offers to play for clubs based a long way from home. In Brazil the major clubs are indeed far away – across the ocean in Europe – and Dona Miguelina might have been wise (if not romantically sentimental) in taking this course of action.

A similar tale of great support been bestowed throughout the formative years is told by England cricketing legend Ian Botham. '*[My father] was always available if I wanted to play cricket... he'd bowl to me endlessly... I had [my parents'] whole-hearted support and encouragement in anything I did, and that was all I needed. The time we spent together and the sacrifices that they made for me are things I remember to this day with pride and gratitude*'.[28]

The youngest heavyweight champion in history (at the tender age of just 20 years and 4 months) Mike Tyson is another striking example of a prodigious athlete who peaked under the guidance of a coach; one he came to consider his father – Cus D'Amato. Tyson was a rough street kid who had been in and out of juvenile hall a number of times by the time D'Amato met him. A phone call from one of the guards at the young Tyson's detention centre, who had a long-standing connection to boxing, alerted D'Amato to the prodigious ability of the young tearaway. D'Amato quickly saw the potential of the teenage Tyson and took him under his wing, inviting him to live and train in his mansion home in Catskill, New York, with other similar surrogates. For a confused and disturbed teenage boy, whose mother died when he was 16 and father was an absentee, this was the ideal environment to call home, '*I was madly in love with Cus D'Amato. He broke me down and built me back up… I wanted him to be my father.*'[29]

However, despite Cus's importance in the eyes of Tyson, the support of D'Amato has also been held up to question. Indeed, world-class trainer Teddy Atlas, who was the principal second behind D'Amato in the early days of Tyson's development, has painted a more cynical side to D'Amato's style of guidance. Tyson, who reportedly had been sexually harassing girls at his high school, was finally expelled after his many misdemeanours were no longer tolerable. But rather than admonishing this behaviour, D'Amato has been said to have brushed such affairs under the carpet, so as not to upset the boxing. There was, for example, no ban from training until his behaviour improved, or something of the like. Atlas has commented that '*Cus… was showing him he was still going to get away with things that someone else couldn't get away with… because [he could] punch, because [he] had the ability to be a world champion fighter… Cus was playing a dangerous game, I thought.*'[30]

If this is indeed true, one can only wonder what kind of life and career Tyson might have had if the pivotal relationship between him and D'Amato had been more conducive to his development as a man as well as a fighter. In many ways, Atlas's observations hint towards much of Tyson's later psychological frailties (although many were also likely to have been present early on). Indeed, while D'Amato might have been a good stabilising influence on an otherwise wayward teen, he might have neglected to provide a moral education for Tyson, in favour of his boxing training and career.

Conclusion

These biographical snippets taken from the formative years of a number of elite athletes go some way in illustrating the validity of much scientific work in the field of sport psychology and expertise. The elite athletes described in this chapter were indeed exposed to a specific set of environmental influences, alongside having possessed favourable basic tendencies, which appear to have collectively fostered their elite abilities. And what the confines of space did not allow to be documented here were the numerous examples of these qualities being evident in many other elite athletes. In fact, I did not find any striking examples of upbringings very different from those predicted. This blueprint for success – if such a thing can truly exist in a formalised sense – seems to have been present in many elite athletes.

That said, Mike Tyson might be the odd one out; his difficult upbringing is in many ways atypical of what one would expect from an elite athlete's formative years. Tyson seems to have had a degree of the necessary influences – certainly an elite training environment stemming from his time at D'Amato's home, as well as fighting on the streets of Brooklyn. However, much also appears to have been missing, including early parental support – he never knew his father, and his mother died while he was still a teenager. This fragmented psychological and emotional development might well be part of the reason why his many successes were blighted by the losses that seemed to arise through emotional breakdowns; the infamous biting of Evander Holyfield's ear – Holyfield lost a portion of his ear – is a case in point.

With the thoughts of Tyson's self-destructive tendencies in mind, it is perhaps fitting to now leave the stories of elite athletes in search of an answer to a question somewhat different to the one we have strived for over the last seven chapters.

If Atlas was correct in asserting that much of Tyson's moral development was overlooked in favour of an old trainer's wish to win the heavyweight prize, the integrity of sport comes under scrutiny. Would Tyson have been a better person having received a different channel for his teenage misdemeanours?

Maybe yes, maybe no. In either case, the question still remains – do we want society's young placed in a cauldron of psychological pressure simply to stand amid the crowd with their hands raised high? The final chapter explores the question – do we want our children to have the life of an elite athlete, and all that it requires and entails? For those who have already made their minds up – I presume that most will have empathically decided in the affirmative – reserve judgement momentarily as we plot a journey through some of the more troubling aspects of youth sports in search of an answer.

8

don't worry ma (i'm only bleeding)

The opening pages of the first chapter asked a fascinating question – what are the origins of the elite sporting mind? And in making it this far one will have snatched a glimpse at the remarkable and vastly complex development of the human mind, not least that of the elite athlete's psychology. However, despite the enormous interest that such origins prompt in the sport-loving public, a second equally prominent question also demands an answer. But while the explanations to the origins of elite athletes are lapped up by the mass media, the rather contrary question of whether or not the making of a champion is actually such a desirable goal after all, is rather less commonly mooted.

One might retort by suggesting that this is because the question itself is a misnomer. After all, is it not the dream of many young boys and girls (if not of those somewhat older) to emulate their sporting heroes? For what possible reason would one *not* wish to make it to the peaks of the sporting world, with fame, fortune, and glory awaiting those who make it to the mountaintop?

However, a body of literature has emerged from those as diverse as medics, sociologists, human rights campaigners, and psychologists – who have begun to piece together a rather different version of events. While the riches and acclaim are certainly extraordinary for those special select few, the margin for error in making it to this higher echelon is so extremely slim.

And the sacrifices needed to achieve elite levels are now so very high that there is a moral imperative to at least consider the implications of a career in professional sport, both for those who might just make it, as well as those who will not.

Chapter 6, 'The Hot House Kids', illustrated the negative psychological implications that can emerge when the enormous pressures to succeed are placed upon the shoulders of young athletes. And this seems to be ever increasing for sportsmen and sportswomen in modern society. The resulting burnout, which a number of young athletes suffer, is a very real phenomenon. However, although certain dangers in the development of young people were alluded to, the emphasis of the chapter centred upon methods of alleviating burnout, with particular regard to maximising potential (rather than for psychological health reasons). And this of course conveniently glosses over the many sensitive issues that elite youth sports provoke.

One of the more well-known concerns of young athletes who commit to these kinds of gruelling training schedules is a loss of a formal childhood experience – socialising with one's peers, developing one's emotional intelligence outside the glare of public criticism, and receiving a solid education. Indeed, while one can modify training regimens so that young athletes are able to maintain their motivation and desire – which in principle is certainly a sound *modus operandi* – this does not mean that training for several hours a day is a wise practice for fostering the child's development outside the sporting world. Engaging with others in a number of settings allows children to learn the many skills, both social and educational, which they fundamentally require in order to become happy and well-adjusted members of society. The question at this juncture is whether the drive to become an elite athlete is compatible with developing such skills. Looking at the evidence it doesn't look like the two are always compatible.

School's out

The spoils of professional sport are convincing more and more families of young athletes to heavily invest their time in cultivating a sporting star, in place of other life skills. In sports, such as tennis, this trend has been particularly extraordinary. Hundreds of young protégés each year now arrive from around the globe to the academies in glamorous locations such as Barcelona and Florida to ply their trade with other similarly ambitious young Turks. British number one Andy Murray was one who left his home in Scotland at 15 to train at the Sanchez-Casal academy in Barcelona.[1] And while the experience of a new culture and peer camaraderie might well be a lifelong highlight, the heavy emphasis upon all things sporting perhaps inevitably involves neglecting, for one, the education of such children.

Indeed, whereas a college education (via a highly sought after full-scholarship) was once seen as the standard route for talented American junior athletes prior to life on the professional tour, this view is no longer upheld. Instead, many now leave college before graduation; the lure of the professional game being too much to resist. The tennis world's number one doubles pair Mike and Bob Bryan commented, '*After winning the NCAA championship, agents offer you high-paying contracts and you don't know if you will get the same opportunity next year. Mike and I decided to join the tour and keep improving our games.*'[2]

In basketball Kobe Bryant sparked a nationwide debate in the United States when he opted to go straight to the NBA and forsake his college studies completely. Bryant's move was perhaps all the more noteworthy because he was by all accounts a strong student who had no need to fear the academic requirements of college life.[3] Bryant's belief in himself at an early age was reportedly high but the gamble he made when he announced his professional intentions was self-apparent, '*I've decided to skip college and take my talent to the NBA ... I don't know if I can reach the stars or the moon. If I fall off a cliff, so be it.'*[4]

This situation in itself is perhaps not overly concerning. After all, many are not cut out for student life and are perfectly well prepared for the job market without the need for letters after their name. However, ever more frequently young athletes are not even making it far enough to make this decision. Tennis champions Pete Sampras, Michael Chang, and Andre Agassi are among those who turned pro while still in their mid-teens, at the tender age of 15 in the case of Chang. This of course meant that even the high school education of these athletes was curtailed.

The same has been observed in many other sports where the pressure to commit wholly to one's athletic dream shunts any hopes for an education to the backburner. And while this is a gamble that Sampras, Chang, and Agassi will certainly not regret, there are countless others who have made the same decisions but found themselves struggling, not only to establish a career in professional sport, but also later outside in the 'real' world. The National Collegiate Athletic Association (NCAA) – the organising body of US collegiate sports – has published data[5] that illustrates just how few make it to the elite echelons that are the pros. In men's basketball just 2.9 per cent of elite US high school players make it to the level of achieving a college scholarship. And of these fortunate few, just 1.3 per cent of athletes are able to make the transition to the NBA. The same is true for a number of other sports.

Aiming for the stars is nothing if not admirable. However, the problem with this ambition stems from its implausibility. Even for those who set the junior sporting world on fire through capturing the most prestigious of prizes, they are by no means guaranteed a successful professional career. A quick glimpse at some of the past junior Wimbledon champions (1992–2003 – those more recent winners haven't had a chance to establish themselves yet so weren't included) shows that half did not even make it to the world's top 100 in subsequent years; the benchmark by which a fledgling tennis professional's success is judged. The stakes are surely high and although pursuing one's dreams is one of the great thrills of life, young athletes deserve an informed choice to the risks they face to their future if they should emerge from their teens with no qualifications and a stalling sporting career.

Innocence lost

While a neglected education is of concern for elite young athletes, the social drawbacks of such a life are also worrying. For those who spend many hours a day perfecting their backhands, back flips, or left hooks, the opportunity to develop social relationships with peers is often highly restricted. World champion boxer Oscar De La Hoya has spoken about the implications of the intense regime his father, Joel, set for him when he was a mere fledgling pugilist in East Los Angeles – '*I had no childhood. I sometimes feel like I missed out on all the activities that all the normal kids used to do.*'[6] One-time darling of the gymnastics world Dominique Moceanu has remarked of her childhood and parents – '*It was always about the gym. I would think: "Don't you guys know anything besides gymnastics? Can't we go for ice cream? Can't you be my mom and dad."*'[7] And seven-time Wimbledon champion Pete Sampras has spoken of a childhood without teen parties and sleepovers, '*my adolescence was pretty much from school to the tennis court, to playing junior tournaments on the weekend. [I] never went to a high school dance or high school football game*'.[8]

While enduring certain hardships and sacrifices in order to achieve one's goals is often viewed somewhat as a right of passage, neglecting the fundamental importance of a child's need to grow as an individual is perhaps a step too far. Indeed, Moceanu infamously 'divorced' her parents when she was 17 having become exhausted at the relentless training schedule she was subjected to.[9] Even more worrying are the cases where young athletes have been worked so hard that they simply suffer complete emotional breakdowns. The unfortunate case of gymnast Christy Henrich is one such example.[10] Henrich had been a US Olympic team hopeful but developed an eating disorder that caused her to plummet in weight to just 47lbs from her original competitive weight of 90lbs (Henrich was just 4'11"). The cause of Henrich's eating disorder has been said to have followed a gymnastic official commenting she needed to lose more weight to be a serious competitor. Tragically Henrich did not recover from her illness; she suffered multiple organ failure and died in hospital at the age of just 22. Her training schedule had reportedly consisted of 20 hours a week from the age of 7 to nine hours a day at the age of 13 while she still attended school.[11]

One must of course take such stories in the appropriate context. Most who are involved in competitive youth sports do not ever approach this kind of commitment, yet reach very respectable levels within their chosen sport. Others willingly take on the sacrifices – long hours of training and limited non-sporting peer interaction – but enjoy every minute of the whole experience. Indeed, although Pete Sampras was quoted above as having had a tennis-centred childhood there is no obvious sentiment that this was anything other than a project of his own initiative.

But the warning signs are not for such individuals. Rather, they have been highlighted to illustrate that extraordinary pressures do indeed exist for a minority of young athletes. And while measures can be taken to stop burnout occurring, if, as some argue, it is true that lengthy hours of gruelling commitment are the necessary prerequisites for elite sporting success, society might be forced to make a difficult decision. Indeed, in the UK the clamour for a home-grown Wimbledon tennis champion has reached almost unprecedented heights. Success at any cost?

Drugs are for winners?

The most common detractions to the elite sporting career were detailed above; the loss of education and a healthy social and emotional development are typical dangers. However, the pressures to maintain one' status at the helm of the sport can lead an athlete to desperate measures. Accordingly, the temptation to seek out the advantages of drugs such as erythropoietin (EPO) and anabolic steroids have never been more apparent, particularly in sports like cycling and sprinting where the finest margins between athletes can be so pivotal.

This course of action has of course been repeatedly splashed all over the sports pages of the daily papers on the many occasions a leading athlete has been caught with such drugs in his or her system. The infamous BALCO case,[12] where the systematic supply of designer steroids to a number of world-class athletes was exposed, has highlighted the depth of such activity. Olympic gold medal winning sprinter Marion Jones was one such disgraced athlete, as was the British sprinter Dwain Chambers.

The reasons for using drugs to create an artificial advantage can easily be explained away as a result of sheer arrogance and lack of regard for the rules of the sport. However, the less cynical critic will perhaps note the comments of athletes such as Chambers who, when asked if drug-free athletes can succeed at the highest level, claimed, *'It's possible, but the person that's taken drugs has to be having a real bad day. That's what I believe.'*[13] The former 100m world record holder Asafa Powell has supported this assessment by Chambers – *'I agree with that 100%'*,[14] although it is important to state that Powell has never tested positive for drugs himself and has added that '*I don't believe in drugs and I don't support anyone that's taking drugs.*'[15]

The suggestion from these comments by athletes who have reached the very helm of their sport is indicative of the widely held sentiment that one cannot compete to serious effect without resorting to such tactics. And after making such tremendous sacrifices throughout one's career to even stand close to the peak, it is perhaps unsurprising that many are overcome by the temptations that exist to cheat. Disgraced Tour de France cyclist Richard Virenque, who tested positive for EPO in 2000, commented '*with my level of education I had no choice: either I made a career in sports or I would have earned the minimum wage for the rest of my life*'.[16] This statement by Virenque, of course, alerts one to the vicious cycle that may exist for many athletes who have placed all their eggs in a single basket – abandoning educational opportunities, and so on – and are thus compelled to travel down a rocky path.

Even more alarming has been the recent finding that many adolescents, and even children as young as ten years old, have been reported to be using performance-enhancing drugs in order to improve their sporting abilities[17] (the figure reported is between 1 and 2 per cent for those in their early teens). For many this is an astonishing statement of affairs – '*could our children really be getting their hands on performance-enhancing drugs?*'. But the findings have been substantiated in further studies alerting one to a more systematic problem. The pressures that are being placed upon our children to succeed in sports appear to be a causal factor. Indeed, it was observed that rather than sport being a protective factor against children experimenting with drugs, the opposite appears to be the case.

The career-ending positive drugs test is one detrimental outcome of such abuse for the athlete. However, the related medical complaints are somewhat more alarming. Drugs such as EPO have been linked to heart problems in those who use them, alongside increased risks of stroke and certain types of cancers. And in recent times a large number of cyclists – in one period, eight in little over a year[18] – have suffered unexpected deaths, largely while still in their sporting prime. Perhaps most infamously, the untimely death of Olympic gold medallist and 100 m and 200 m record holder Florence Griffiths-Joyner was greeted with scepticism by those who felt her extraordinary successes could only have been drug-assisted. Indeed, she improved her 100 m time by an unprecedented several tenths of a second in the season of her Olympic and world-record breaking successes. One can only wonder how many ticking time bombs exist out there.

In fairness to Griffiths-Joyner and many of the cyclists who have passed away prematurely, the jury is still out with regard to the role of drugs abuse in their deaths. Griffiths-Joyner was reported as having suffered an epileptic seizure – which in turn led to death by asphyxiation – rather than the heart attack the press had initially suggested.[19] And similar inconsistencies in the apparent cause of death have been noted in the various cycling fatalities as well.[20] But this of course does not reduce the very real danger of abusing drugs. As Virenque has noted, sometimes the pressure to succeed becomes so great the risks of ill-health and public disgrace appear to be ones worth taking.

The social death of a sportsman

This rather dramatic heading may appear at first to be a strange phrase to use for a professional athlete. After all, the social status of elite athletes in the current age is little short of sky-high. However, the 'social death' referred to is less an observation of life while an active competitor, but rather the challenges that arrive following one's retirement. For while the average retirement age of the working population is around 65, the retiring athlete is somewhat closer to 30 when he or she hangs up their boots. And this social anomaly has been noted as an especially difficult transition in the life of a professional athlete. Indeed, whereas many of the athlete's peers will be on their umpteenth promotion and settled into a career path, the professional athlete suddenly find themselves in the proverbial wilderness, in need of a new direction after so many years cultivating a singular ambition.

For some this sentiment is an absurdity. After all, aren't elite athletes rather well provisioned for, and able to retire to a comfortable existence with the wife and kids? This view has garnered support in the literature[21] with suggestions that the contacts that ex-athletes build up over their professional sporting years enable them to make the transition into non-sporting professional pursuits with greater ease than is often acknowledged; a degree of notoriety as a professional athlete is suggested to work wonders at job interviews. This is in addition to the fact that many athletes will have become tired of the relentless training regimes that are so central to the life of an elite performer. As such, they may well welcome the change in lifestyle and the chance to settle into a more sedate pace of life.

However, problems do seem to be arising, despite the apparently favourable life opportunities that often exist for ex-professional athletes. For one, consider the statistic that reports almost 80 per cent of American footballers in recent retirement from the NFL were bankrupt, unemployed, and/or divorced.[22] Former offensive lineman for the Greenbay Packers Ken Ruettgers has said of retirement, '*I had played football for two-thirds of my life. It was my passion and purpose. It's what I was made to do. And then, it was over... No more autographs, applause, million-dollar contracts, front-page photos, and product endorsements. Retirement is more than the end of a job. It feels like death.*'[23]

And it is this aspect of life that many who have never been privy (myself included) will probably never be able to understand. For those who have tasted the adulation of millions, been recognised on the street, and had marriage proposals slid under hotel doors, the step-down into normal life is undoubtedly a hard one. This is perhaps most prevalent in boxing, where the aging pugilist all-too often simply doesn't seem to know when to call it a day. The notion that a boxer should choose to retire from the sport, rather than the sport finally retiring the boxer, is an old adage that certainly proffers sound advice but sadly often goes unheeded. Four-time world heavyweight champion Evander Holyfield – who famously tamed Mike Tyson – is the latest in a long line of fighters who have thrown caution (and medical wisdom) to the wind and continued to fight. Holyfield's inspiration to fight well beyond his fortieth birthday is to retire as the heavyweight champion, claiming in 2007 – '*the best is yet to come. I'm not going to retire until I'm the undisputed heavyweight champion of the world again*'.[24] Yet some have sounded the warning signs that all is not alright with Holyfield, suggesting that the early signs of pugilistica dementia – punch-drunk syndrome – are evident in his speech.[25]

While Holyfield still plies his trade despite the views of his detractors, the much-loved former world heavyweight champion, Frank Bruno, has endured perhaps a more torrid time than he ever experienced in his brace of ring wars with Mike Tyson. Bruno suffered a well-documented mental breakdown in recent years that necessitated professional intervention, an outcome that some have put down to his retirement 'blues'. As Henry Cooper, former British heavyweight champion, has stated, '*If you don't have that daily routine any more, where you have to go to the gym at a certain time, train at a certain time, you miss it – all fighters do… One day you are a boxer and the next you wake up and you have retired – you are not a boxer, then you think, "what am I going to do now?"'*.[26] And Bruno is perhaps just one of a number who are left facing an uncertain life following their retirement from sport.

It is worth reiterating once again that there is no case for claiming that the hot house academies, and ambitions to be an elite athlete are necessarily a negative enterprise in themselves. For many youngsters the thought of being able to indulge oneself in the sport that one loves for as long as there are waking hours in a day is surely a dream come true. What parent would not support a child's ambition, particularly when a successful and lucrative career can be formed off the back of a passion? However, this argument can also be used in reverse. For although children may love computer games and candy, no self-respecting parent or guardian would allow their offspring to spend hours every day playing Mario Bros. or consuming chocolate bar after chocolate bar. The role of the parent or guardian is as much to know when to hold the child back from too much of a good thing as to when to push the child to strive for just that one more step when they are otherwise unmotivated themselves. The solution appears, perhaps as always, to keep a sense of what is best for all concerned, and not as we have seen in the case of Christy Henrich, allow a child to be completely overwhelmed with competitive pressures to the point that his or her very life is jeopardised.

Final conclusion

We have come a considerable way since the challenge set in the opening chapter to explain the origins of the elite sporting mind. So, what have we learnt along the way? What factors underpin the championship psychology of a Wimbledon champion, an England batsman, or a British Lions scrumhalf?

As one will now be only too aware, there is no quick answer to such a question. Certainly, the notion of an elite sporting psychology solely emerging from innate gifts, or, for that matter, from sheer hard graft is a gross misunderstanding of the mind's development. The psychology of Roger Federer, Tiger Woods, Ronaldinho, as well as you and I, are all the culmination of an extraordinary interplay between genes and environment that science is only just beginning to understand with any success.

The elite sporting personality traits – high drive, diligence, motivation, and self-reliance, among others – were shown to hold a significant genetic influence; indeed, as much as 50 per cent of our personality seems to arise from the lottery that is one's genes. And so, for every sporting parent, there is a good chance that a sporting child will follow.

But, this of course leaves 50 per cent of one's personality unaccounted for. To this end, we saw a number of the factors that have been hypothesised to sculpt one's personality. The conditions of the womb, the nature of the early interactions (typically) with one's mother, and the dynamics within the family – birth order effects – all featured as likely suspects in this process. No doubt, they all do play some role. However, the influence of the peer group in defining one's personality is perhaps the most profound of them all as the evidence from the twin studies illustrate. With this in mind, the fundamental importance of an elite peer group/set of role models is only too apparent. To have a chance at making it to the top of the sporting world, an aspiring young athlete will need to model themselves upon those who are already enjoying the early fruits of success. The importance of elite training squads is at least as important for establishing the values and traits of a young protégé, as much as it is to foster their fitness and technique.

Of course, the elite training setting does not only offer this facility. As we have seen with the development of a 'magic eye', great cognitive skill is evident in the perceptual and decision-making processes of all elite performers. And for the most part, this skill is a learnable feature of the elite athlete's repertoire given the right environment. The right environment appears to be one that fosters a wide range of tactical plays, so that the young hopeful can experience a wide range of scenarios and begin to compile the appropriate responses that become so effortless for those special few.

However, while much emphasis has been placed on the role of nurture in perceptual and decision-making expertise, it is important to note that one's innate influences in this domain cannot be forgotten. Indeed, great individual differences are known to exist in the ability to construct and manipulate patterns in the mind's eye. This faculty is known to be a highly heritable trait and so the ability to learn elite perceptual skills is almost certainly going to be mediated by the degree of one's innate ability to detect patterns among apparent chaos.

The faculty of mental toughness is also comprised of both nature and nurture. Cognitive skills, such as self-talk and visual imagery, are learned techniques commonly used by elite athletes to control their emotional arousal, allowing them to stay in control of their performance. Carol Dweck has also emphasised that whether one adopts a theory of 'fixed' or 'malleable' ability will prompt different coping strategies in times of strife.

Yet despite the skills that sport psychologists now recommend to bolster one's mental toughness, there are also innate factors at work. The executive functions – the metaphorical 'conductor' at work inside the brain – have been shown to be highly heritable. This of course suggests that the ability to block out distractions – what psychologists term as response inhibition – is not a skill that all can necessarily master, despite the numerous strategies that are available to the athlete.

When all is said and done, the elite sporting mind is nothing if not a mixing-pot of many ingredients. And we have discovered many of these factors crucial to the making of a sporting champion in this book, although admittedly there are likely to be more still hidden from the probing eyes of science. Certainly, we are still some way from truly understanding the mind of an elite champion. Nonetheless, we have come a long way in deciphering the origins of the elite sporting mind in recent decades. No longer are the extraordinary reactions of the world champion boxer, or the mental toughness of the hardy Olympic marathoner, the mysterious powers they were once viewed to be. Only time, of course, can tell what further advances sport science will make in comprehending the making of a champion's mind. However, if they are anything like the understandings made in recent times they will be worth looking out for.

notes

Chapter 1

1. K. Jones, Ali v. Frazier–'It was like death. Closest thing to dyin' that I know of.' The Independent, September 30 2005. (http://www.independent.co.uk/sport/general/boxing-ali-v-frazier—it-was-like-death-closest-thing-to-dyin-that-i-know-of-508934.html). Retrieved 04/06/08.
2. See (http://thinkexist.com/quotation/baseball_is_ninety_percent_mental_and_the_other/226268.html). Retrieved 04/06/08.
3. Kirby, G. Pele, King of Futbol. ESPN.com, 2007. (http://espn.go.com/classic/biography/s/Pele.html) Retrieved 04/06/08.
4. Oscar de la Hoya: Beyond the glory [documentary], Fox Sports Net.
5. A. M. Williams, K. Davids, & J. G. Williams (1999) *Visual perception and action in sport*, Spon Press.
6. B. Saltin, H. Larsen, N. Terrados, J. Bangsbo, T. Bak, C. K. Kim, et al. (1995) Aerobic exercise capacity at sea level and at altitude in Kenyan boys, junior and senior runners compared with Scandinavian runners, *Scandinavian Journal of Medicine & Science in Sports*, 5(4), 209–221.
7. G. Kolata, Super, Sure, but Not More Than Human. New York Times, July 24, 2005. (http://www.nytimes.com/2005/07/24/weekinreview/24kola.html). Retrieved 04/06/08.
8. T. Noakes (1991) *Lore of running*, Leisure Press.
9. A. G. Scrimgeour, T. D. Noakes, B. Adams, & K. Myburgh (1986) The influence of weekly training distance on fractional utilization of maximum aerobic capacity in marathon and ultramarathon runners, *European Journal of Applied Physiology*, 55, 202–209.
10. J. Entine, Sport and ethnicity (http://www.pponline.co.uk/encyc/0657b.htm). Retrieved 04/06/08.
11. Y. P. Pitsiladis, V. O. Onywera, E. Geogiades, W. O'Connell, & M. K. Boit (2004) The dominance of Kenyans in distance running, *Equine and Comparative Exercise Physiology*, 1(4), 285–291.
12. Mike Tyson: Beyond the glory [documentary], Fox Sports Net.
13. J. Gascoigne (2008) *Captain Cook: Voyager between worlds*, Hambledon Continuum.

Chapter 2

1. M. Ridley (2004) *Nature via nurture: Genes, experience, and what makes us human*, Harper Perennial.
2. J. D. Watson (2004) The double helix: A Personal account of the discovery of the structure of DNA, Harper Perennial.
3. A. Coghlan, Gene variant linked to athletic performance, New Scientist, August 27 2003 (http://www.newscientist.com/article/dn4092-gene-variant-linked-to-athletic-performance.html). Retrieved – 04/06/08.
4. R. Dawkins (1976) *The selfish gene*, Oxford University Press.
5. E. Pennisi (2007) Working the (gene count) numbers: Finally, a firm answer? *Science*, 316(5828), 1113.
6. E. Pennisi (2007) Sea anemone provides a new view of animal evolution, *Science*, 317(5834), 27.
7. K. Carr (1993) Nobel goes to discoverers of 'split genes', *Nature*, 365, 597.
8. R. Lickliter & H. Honeycutt (2003) Developmental dynamics: toward a biologically plausible evolutionary psychology, *Psychological Bulletin*, 129(6), 819–835.
9. R. Dawkins (1985) Review of – Not in our genes: Biology, ideology and human nature by Steven Rose, Leon J. Kamin and R. C. Lewontin. From "Sociobiology: the debate continues', *New Scientist*, 24 January.
10. P. Morgan How the Giants kicked tradition into touch with psychometric testing, peak performance Online. (http://www.pponline.co.uk/encyc/0806.htm) Retrieved – 04/06/08; Sando, M. Leaf's San Diego failure had ripple effect, ESPN.com, April 9, 2008 (http://sports.espn.go.com/nfl/draft08/columns/story?columnist=sando_mike&id=3335873). Retrieved – 04/06/08.
11. L. Festinger, H. W. Riecken, & S. Schachter (1956) *When prophecy fails: A social and psychological study of a modern group that predicted the end of the world*, University of Minnesota Press.
12. R. R. McCrae et al. (2000) Nature over nurture: Temperament, personality, and lifespan development, *Journal of Personality and Social Psychology*, 78(1), 173–186.
13. Matthew XXV:29, King James Version.
14. K. E. Stanovich (1986) Matthew effects in reading: Some consequences of individual differences in the acquisition of literacy, *Reading Research Quarterly*, 21(4), 360–407.
15. D. Remnick (1999) *King of the world*, Picador.
16. N. Bollettieri & C. H. Maher (1996) *Nick Bollettieri's mental efficiency program for playing great tennis*, Contemporary Books Inc.
17. S. Carp, Trainer a father figure to Pacquiao. Las Vegas Review. March 14, 2008. (http://www.lvrj.com/sports/16671586.html). Retrieved – 04/06/08.

Chapter 3

1. D. James, If you thought trainspotters were weird ... try footballers, guardian.co.uk, October 22, 2006. (http://blogs.guardian.co.uk/sport/2006/10/22/if_you_thought_trainspotters_w.html). Retrieved – 04/06/08.
2. L. Cooper (1969) Athletics, activity, and personality: A review of the literature, *Research Quarterly*, 40(1), 17–22.
3. E. V. Aidman (2007) Attribute-based selection for success: The role of personality attributes in long-term predictions of achievement in sport, *The Journal of the American Board of Sport Psychology*, 1, article 3.
4. L. R. Goldberg (1993) The structure of phenotypic personality traits, *American Psychologist*, 48(1), 26–34.
5. S. J. McKelvie, P. Lemieux, & D. Stout (2003) Extraversion and neuroticism in contact athletes, no contact athletes and non-athletes: A research note, Athletic Insight, *The Online Journal of Sport Psychology*, 5(3).
6. M. Johnson (1996) *Slaying the dragon: How to turn your small steps to great feats*, Harper Collins.
7. M. Zuckerman (1995) Good and bad humors: Biochemical bases of personality and its disorders, *Psychological Science*, 6(6), 325–332.
8. D. Nettle (2007) *Personality: What makes you the way you are*, Oxford University Press. p. 92.
9. M. B. Stein, A. N. Simmons, J. S. Feinstein, & M. P. Paulus (2007) Increased Amygdala and Insula activation during emotion processing in anxiety-prone subjects, *American Journal of Psychiatry*, 164, 318–327.
10. A. Bechara, A. R. Damasio, H. Damasio, & S. W. Anderson (1994) Insensitivity to future consequences following damage to human prefrontal cortex, *Cognition*, 50, 7–15.
11. J. C. Loehlin, R. R. McCrae, P. T. Costa, & O. P. John (1998) Heritabilities of common and measure-specific components of the big five personality factors, *Journal of Research in Personality*, 32, 431–453.
12. A. J. DeCasper & M. J. Spence (1986) Prenatal maternal speech influences newborns' perception of speech sound, *Infant Behaviour and Development*, 9, 133–150.
13. L. K. Takahashi, J. G. Turner, & N. H. Kalin (1991) Prenatal stress alters brain catecholaminergic activity and potentiates stress-induced behaviour in adult rats, *Brain Research*, 574, 131–137.
14. B. M. Gutteling, C. Weerth, S. H. Willemsen-Swinkels, A. C. Huizink, E. J. Mulder, G. H. Visser, & J. K. Buitelaar (2005) The effects of prenatal stress on temperament and problem behavior of 27-month-old toddlers, *European Child and Adolescent Psychiatry*, 14, 41–51.
15. J. Van Os & J. P. Selten (1998) Prenatal exposure to maternal stress and subsequent schizophrenia: the May 1940 invasion of the Netherlands, *British Journal of Psychiatry*, 172, 324–326.

16. T. B. Brazelton, E. Tronick, L. Adamson, H. Als, & S. Wise (1975) Early mother-infant reciprocity, Ciba Foundation Symposium 33, Amsterdam, Elsevier.
17. E. Fivaz-Depeursinge (1991) Documenting a time-bound, circular view of hierarchies: A microanalysis of parent-infant dyadic interaction, *Family Process*, 30, 101–120.
18. R. Rosenthal & L. Jacobson (1992) *Pygmalion in the classroom*, New York: Irvington.
19. T. Sternberg Horn, C. L. Lox, & F. Labrador (2001) The self-fulfilling prophecy theory: When coaches' expectations become reality. In J. M. Williams (ed.) (2001) Applied sport psychology: Personal growth to peak performance, Mayfield Publishing.
20. M. Jordan & T. Hatfield (2005) *Driven from within*, Simon & Schuster Ltd.
21. P. Kristensen & T. Bjerkedal (2007) Explaining the relation between birth order and intelligence, *Science*, 316, 1717.
22. S. E. Black, P. J. Devereux, & K. J. Salvanes (2005) The more the merrier? The effect of family size and birth order children's education, *The Quarterly Journal of Economics*, 120(2), 669–700.
23. R. D. Clark & G. A. Rice (1982) Family constellations and eminence: The birth orders of Nobel prize winners, *Journal of Psychology*, 110, 281–287.
24. F. J. Sulloway (1997) *Born to rebel: Birth order, family dynamics, and creative lives*, Vintage Books.
25. Those who equate scientific achievement with creativity and non-conformity might be surprised to find later-borns winning fewer Nobel prizes. Clark and Rice (1982) have suggested that the early prizes reflected more conservative breakthroughs, whereas the later prizes have been more markedly innovative. With this in mind, they have noted greater later-born Nobel laureates in recent years.
26. R. B. Zajonc, H. Markus, & G. B. Markus (1979) The birth order puzzle, *Journal of Personality and Social Psychology*, 37(8), 1325–1341.
27. Ibid.
28. E. Woods (1997) *Training a tiger; the official book on how to be the best*, Hodder & Stoughton, p. xvi.
29. J. R. Harris (2006) *No two alike: Human nature and human individuality*, W.W. Norton & Co.
30. L. Kanner (1949) Problems of nosology and psychodynamics in early childhood autism, *American Journal of Orthopsychiatry*, 19(3), 416–426.
31. J. R. Harris (1998) *The nurture assumption: Why children turn out the way they do*, Bloomsbury.
32. R. Plomin & D. Daniels (1987) Why are children in the same family so different from each other? *Behavioral and Brain Sciences*, 10, 1–16.
33. Ibid.
34. A. Bandura (1977) Self-efficacy: Towards a unifying theory of behavioural change, *Psychological Review*, 84(2), 191–215.

35. Quoted in Sampras Builds More Success on Top of Confidence, The New York Times, November 18, 1991. (http://query.nytimes.com/gst/fullpage.html?res=9D0CE1D71E3FF93BA25752C1A967958260). Retrieved – 04/06/08.
36. See (http://www.quotedb.com/quotes/2418). Retrieved – 04/06/08.
37. Quoted in The boy who fell to Earth, Tennis Magazine, April 2001. (http://www.toure.com/CONTENT/ARTICLES/alparker.htm). Retrieved – 04/06/08.
38. Ibid.

Chapter 4

1. J. Stevenson, How will English football develop? British Broadcasting Corporation. (http://news.bbc.co.uk/sport1/hi/football/7137071.stm). Retrieved – 04/06/08.
2. R. Dawkins (1986) *The blind watchmaker*, New York: W. W. Norton & Company.
3. Ibid.
4. R. English (1915) Democritus' theory of sense perception, Transactions and Proceedings of the American Philological Association, 46, 217–227.
5. C. Fullerton (1925) Eye, ear, brain, and muscle tests on Babe Ruth, Western Optometry World 13(4), 160–1.
6. A. Sherman (1980) Overview of research information regarding vision and sports, *Journal of American Optometric Association*, 51, 661–666.
7. J. Baker, S. Horton, J. Robertson-Wilson, & M. Wall (2003) Nurturing sport expertise: Factors influencing the development of elite athletes, *Journal of Sports Science and Medicine*, 2, 1–9.
8. A. M. Williams, K. Davids, & J. G. Williams (1999) *Visual perception and action in sport*, Spon Press (p. 84).
9. M. F. Land & P. McLeod (2000) From eye movements to actions: how batsmen hit the ball, *Nature Neuroscience*, 3, 1340–1345.
10. Ibid.
11. G. A. Miller (1956) The magical number seven, plus or minus two. *The Psychological Review*, 63(2), 81–97.
12. N. Cowan (2001) The magical number 4 in short-term memory: A reconsideration of mental storage capacity, *Behavioral and Brain Sciences*, 24, 87–185.
13. W. G. Chase & H. A. Simon (1973) Perception in chess. *Cognitive Psychology*, 4, 55–81.
14. Ibid.
15. L. Schwartz, 'Great' and 'Gretzky' belong together, ESPN.com. (http://espn.go.com/sportscentury/features/00014218.html). Retrieved – 04/06/08.

16. C. Jardine, Sir Jackie Stewart: Winning is not enough, The Daily Telegraph, September 29, 2007. (http://www.telegraph.co.uk/portal/main.jhtml?xml=/portal/2007/09/29/nosplit/ftfront.xml), Retrieved – 04/06/08.
17. C. M. Jones & T. R Miles (1978) Use of advance cues in predicting the flight of a lawn tennis ball, *Journal of Human Movement Studies*, 4, 231–235.
18. S. L. McPherson & M. W. Kernodel (2003) Tactics, the neglected attribute of expertise: Problem representations and performance skills in tennis. In J. L. Starkes & K. Anders Ericsson (2003) *Expert Performances in Sport: Advances in Research on Sport Expertise*, Human Kinetics.
19. Ibid.
20. Home Box Office (April 2008) Countdown to Hopkins-Calzaghe [documentary].
21. L. Schwartz, 'Great' and 'Gretzky' belong together, ESPN.com. (http://espn.go.com/sportscentury/features/00014218.html). Retrieved – 04/06/08.
22. D. Farrow & M. Raab (2008) A recipe for expert decision making. In D. Farrow, J. Baker, & C. MacMahon (2008) *Developing sport expertise: Researchers and coaches put theory into practice*, Routledge.
23. D. Farrow & B. Abernethy (2002) Can anticipatory skills be learned through implicit video-based perceptual training? *Journal of Sports Sciences*, 20, 471–485.
24. http://news.bbc.co.uk/sport1/hi/tennis/skills/4237736.stm.
25. J. L. Starkes & S. Lindley (1994) Can we hasten expertise by video simulations? *Quest*, 46, 211–222.
26. Ibid.
27. R. Masters (2008) Skill learning the implicit way – say no more! In D. Farrow, J. Baker, & C. MacMahon (2008) *Developing sport expertise: Researchers and coaches put theory into practice*, Routledge.
28. A. Benson & J. Sinnott, Why are Brazil so good? BBC.co.uk, May 23, 2006, (http://news.bbc.co.uk/sport1/hi/football/world_cup_2006/teams/brazil/4751387.stm). Retrieved – 04/06/08.
29. C. B. Mervis, B. R. Robinson, & J. R. Pani (1999) Cognitive and behavioural genetics '99<: Visuospatial construction, *American Journal of Human Genetics*, 65, 1222–1229.
30. Ibid.

Chapter 5

1. See (http://ironman.com/). Retrieved – 04/06/08.
2. J. Moss, The Most Famous Finish in Ironman History: Julie Moss Takes You Through Her Race, Ironman.com. (http://ironman.com/holdingcell/2003/february-2003/the-most-famous-finish-in-ironman-history-julie-moss-takes-you-through-her-race). Retrieved – 04/06/08.

3. See the footage posted on YouTube (http://www.youtube.com/watch?v=5NzMTLMJwao).
4. See note 2 above.
5. G. Jones (2002) What is this thing called mental toughness? An investigation of elite sport performers, *Journal of Applied Sport Psychology*, 14(3), 205–218.
6. Nadal departs as four of top five seeds fall, The Guardian, February 22, 2008. (http://www.guardian.co.uk/sport/2008/feb/22/tennis.sport). Retrieved – 04/06/08.
7. C. Clarey, A spectator disrupts the marathon with a shove, The New York Times, August 30 2004. (http://query.nytimes.com/gst/fullpage.html?res=9A06E3DB1631F933A0575BC0A9629C8B63). Retrieved – 04/06/08.
8. E. Goldberg (2004) *The Executive Brain: Frontal Lobes and the Civilized Mind*, Oxford University Press.
9. Ibid.
10. D. Carlin, J. Bonbera, M. Phipps, G. Alexander, M. Shapiro, & J. Grafman (2000) Planning impairments in frontal lobe dementia and frontal lobe patients, *Neuropsychologia*, 38, 655–665.
11. R. S. Wilson, D. A. Bennett, J. L. Bienias, N. T. Aggarwal, C.F. Mendes de Leon et al. (2002) Cognitive activity and incident AD in a population-based sample of older persons, *Neurology*, 59, 1910–1914.
12. For a good review see G. Lawton, Is it worth going to the mind gym? New Scientist, 12 January 2008.
13. H. W. Mahncke, B. B. Connor, J. Appelman, O. N. Ahsanuddin, J. L. Hardy et al. (2006) Memory enhancement in healthy older adults using a brain plasticity-based training program: A randomized, controlled study, Proceedings of the National Academy of Sciences, 103(33), 12523–12528.
14. D. M. Wegner (1994) Ironic processes of mental control, *Psychological Review*, 101, 34–52.
15. B. Miller (1997) *Gold Minds: the psychology of winning in sports*, The Crowood Press.
16. M. Campbell, A strong mind, BBC.co.uk. (http://news.bbc.co.uk/sport1/hi/golf/skills/4381450.stm). Retrieved 04/06/08.
17. G. Garber, Gamesmanship is name of the game in tennis, ESPN.com. August 9, 2007. (http://sports.espn.go.com/espn/cheat/news/story?id=2955743). Retrieved 04/06/08.
18. D. J. Crews & S. H. Boutcher (1986) Effects of structured pre-shot behaviours on beginning golf performance, *Perceptual and Motor Skills*, 62, 291–294.
19. M. Campbell, A strong mind, BBC.co.uk. (http://news.bbc.co.uk/sport1/hi/golf/skills/4381450.stm). Retrieved 04/06/08.
20. B. Miller (1997) *Gold Minds: the psychology of winning in sports*, The Crowood Press, p. 56.

21. C. Dweck (2000) Self-theories: Their role in motivation, personality, and development (Essays in social psychology), Psychology Press Ltd.
22. M. Jordan (2006) Driven from within, Atria Books, p. 4.
23. C. Dweck (2000) Self-theories: Their role in motivation, personality, and development (Essays in social psychology), Psychology Press Ltd.
24. Ibid.
25. D. Ornstein, My nerves threatened to get better of me, says Wilkinson, The Guardian, February 5 2007. (http://www.guardian.co.uk/sport/2007/feb/05/sixnations2007.sixnations). Retrieved 04/06/08.
26. W. Cannon (1929) *Bodily changes in pain, hunger, fear, and rage*, New York: Appleton.
27. Calzaghe 'can take on the world', BBC.co.uk, April 21 2002. (http://news.bbc.co.uk/sport1/hi/boxing/1942260.stm). Retrieved 04/06/08.
28. B. Miller (1997) *Gold Minds: the psychology of winning in sports*, The Crowood Press, p. 36.
29. England win Rugby World Cup, BBC.co.uk, November 22 2003, (http://news.bbc.co.uk/sport1/hi/rugby_union/international/3228728.stm). Retrieved 04/06/08.
30. D. Lavallee, J. Kremer, A. P. Moran, & M. Williams (2004) *Sport psychology: Contemporary themes*, Palgrave Macmillan.
31. G. Ganis, W. L. Thompson, & S. M. Kosslyn (2004) Brain areas underlying visual mental imagery and visual perception: an fMRI study, *Cognitive Brain Research*, 20, 226–241.
32. B. Miller (1997) *Gold Minds: the psychology of winning in sports*, The Crowood Press, p. 35.
33. Oscar De La Hoya: Beyond the Glory [documentary] Fox Sports Net.
34. See (http://www.selfgrowth.com/quote_2003.html).
35. J. Kremer & D. M. Scully (1994) *Psychology in sport*, Taylor & Francis, p. 75.
36. M. J. Mahoney & M. Avener (1977) Psychology of the elite athlete: An exploratory study, *Cognitive Therapy and Research*, 1, 135–141. D. Gould, R. C. Eklund, & S. Jackson (1992) Coping strategies used by more versus less successful Olympic wrestlers, *Research Quarterly for Exercise and Sport*, 64, 83–93.
37. D. H. Meichenbaum & J. Goodman (1971) Training impulsive children to talk to themselves: A means of developing self-control, *Journal of Abnormal Psychology*, 77(2), 115–126.
38. S. Shetty, The biggest upset ever, BBC.co.uk, January 14 2002. (http://news.bbc.co.uk/sport1/hi/boxing/specials/ali_at_60/1722098.stm). Retrieved 04/06/08.
39. 50 greatest sporting insults, The Times Online, August 1 2007. (http://www.timesonline.co.uk/tol/sport/football/article2178440.ece). Retrieved 04/06/08.

40. M. Campbell, A strong mind, BBC.co.uk. (http://news.bbc.co.uk/sport1/hi/golf/skills/4381450.stm). Retrieved 04/06/08.
41. Eminem tops Premiership warm-up chart, BBC.co.uk, May 8 2003. (http://news.bbc.co.uk/cbbcnews/hi/music/newsid_3011000/3011353.stm). Retrieved 04/06/08.
42. A. M. Lane (2006) Reflections of professional boxing consultancy: A response to Schinke (2004), *Athletic Insight*, 8(3).
43. H. J. Eysenck (1967) *The biological basis of personality*, Springfield, IL: Thomas.
44. G. Belojevic, V. Slepcevic, & B. Jakovljevic (2001) Mental performance in noise: The role of introversion, *Journal of Environmental Psychology*, 21, 209–213.
45. N. P. Friedman, A. Miyake, S. E. Young J.C. Defries, R. P. Corley, & J. K. Hewitt (2008) Individual differences in executive functions are almost entirely genetic in origin, *Journal of Experimental Psychology: General*, 137(2), 201–225.

Chapter 6

1. P. Cartledge (2001) *Spartan reflections*, Gerald Duckworth & Co Ltd.
2. T. Fordyce, What are the Olympic ideals? BBC.co.uk, May 27 2003. (http://news.bbc.co.uk/sport1/hi/other_sports/2941062.stm). Retrieved – 04/06/08.
3. See (http://www.brainyquote.com/quotes/authors/v/vince_lombardi.html).
4. M. McCarthy, Wake up consumers? Nike's brash CEO dares to just do it, USA Today, June 16 2003. (http://www.usatoday.com/money/advertising/2003-06-15-nike_x.htm). Retrieved – 04/06/08.
5. P. Bandini, Beckham confirms LA Galaxy move, guardian.co.uk, January 11 2007. (http://www.guardian.co.uk/football/2007/jan/11/newsstory.europeanfootball). Retrieved – 04/06/08.
6. E. Bradley, Tiger Woods Up Close And Personal, CBSNEWS.com, September 3 2006. (http://www.cbsnews.com/stories/2006/03/23/60minutes/main1433767_page8.shtml). Retrieved – 04/06/08.
7. G. Turner, Premiership salaries up 65% since 2000, The Guardian, April 11 2006. (http://www.guardian.co.uk/football/2006/apr/11/newsstory.sport10). Retrieved – 04/06/08.
8. P. David (2005) *Human rights in youth sports: A critical review of children's rights in competitive sports*, Routledge Publishing. p. 135.
9. Ibid., p. 136.
10. French tennis drug father jailed, BBC.co.uk, March 9 2006. (http://news.bbc.co.uk/1/hi/world/europe/4789066.stm) Retrieved – 04/06/08.

11. J. Howard, Harding Admits Knowing of Plot After the Attack, The Washington Post, January 28 1994. (http://www.washingtonpost.com/wp-srv/sports/longterm/olympics1998/history/timeline/articles/time_012894.htm) Retrieved – 04/06/08.
12. C. Sellers, Working toward perfection: An interview with Martha Karolyi. (http://coaching.usolympicteam.com/coaching/kpub.nsf/v/3karo04). Retrieved – 04/06/08.
13. M. J. A. Howe (1990) *The origin of exceptional abilities*, Basil Blackwell.
14. R. Finn, The Molding of a Tennis Prodigy, The New York Times, April 23 1992. (http://query.nytimes.com/gst/fullpage.html?res=9E0CE5DE1F38F930A15757C0A964958260). Retrieved – 04/06/08.
15. N. Harris, Improving the Bollettieri way: two British youngsters hit the courts in Florida, The Independent, December 20 2005. (http://www.independent.co.uk/sport/tennis/improving-the-bollettieri-way-two-british-youngsters-hit-the-courts-in-florida-520194.html). Retrieved – 04/06/08.
16. L. Ryan, Mary Lou Retton: All About Lou, CBN.com. (http://www.cbn.com/entertainment/sports/ChristianAthletes/marylou_retton.aspx). Retrieved – 04/06/08.
17. Aristotle (1995) *Politics*, Oxford World's Classics, Oxford University Press. p. 303.
18. Ibid.
19. P. Newman, Hingis quits under cocaine cloud, The Independent, November 2 2007. (http://www.independent.co.uk/sport/tennis/hingis-quits-under-cocaine-cloud-398631.html). Retrieved – 04/06/08.
20. Henin announces shock retirement, BBC.co.uk. (http://news.bbc.co.uk/sport1/hi/tennis/7399963.stm). Retrieved – 04/06/08.
21. D. Robson, Professional Tennis at Age 15: Too Much to Young? The Washington Post, April 18 2005. (http://www.washingtonpost.com/wp-dyn/articles/A61484-2005Apr17.html). Retrieved – 04/06/08.
22. P. Wachter, Prodigy's End. The New York Times, June 24, 2007. (http://www.nytimes.com/2007/06/24/magazine/24young-t.html?ei=5090&en=51c38e4b7b87b77d&ex=1340337600&partner=rssuserland&emc=rss&pagewanted=all). Retrieved – 04/06/08.
23. R. J. Rotella, T. Hanson, & R. H. Coop (1991) *Burnout in youth sports*, The Elementary School Journal, 91(5), 421–428.
24. Ibid.
25. R. E. Smith (1986) Toward a cognitive-affective model of athletic burnout, *Journal of Sport Psychology*, 8, 36–50.
26. Tired O'Sullivan considers break, BBC.co.uk, April 27 2005. (http://news.bbc.co.uk/sport1/hi/other_sports/snooker/4492107.stm). Retrieved – 04/06/08.
27. I. Balyi & A. Hamilton (2000) *Key to Success: Long-term Athlete Development.* Sport Coach, Canberra, Australia. 23(1), 30–32.

28. D. Coyle, How to Grow a Super-Athlete, The New York Times, March 4 2007. (http://www.nytimes.com/2007/03/04/sports/playmagazine/04play-talent.html?pagewanted=4&_r=3&ei=5087&em&en=ecf33b897e57fc82&ex=1173330000). Retrieved – 04/06/08.
29. R. Philip (1993) *Agassi: The fall and rise of the enfant terrible of tennis*, Bloomsbury Publishing, p. 46.
30. T. O. Bompa (1983) *Theory and methodology of training: The key to athletic performance*, Kendall/Hunt Publishing Company.
31. S. Nakrani Three months of hard slog, body-belts and eating clean. The Guardian, December 7 2007. (http://www.guardian.co.uk/sport/2007/dec/07/boxing.rickyhatton). Retrieved – 04/06/08.
32. Ibid.
33. Ibid.
34. H. Selye (1936) A syndrome produced by diverse nocuous agents, *Nature*, 138, 32.
35. B. Gilbert, Four kilos and eight mph can make Murray a champion, The Guardian, June 2 2007. (http://blogs.guardian.co.uk/sport/2007/06/02/four_kilos_and_eight_mph_can_m.html). Retrieved – 04/06/08.
36. L. Hardy, G. Jones, & D. Gould (1996) *Understanding psychological preparation for sport: Theory and practice of elite performers*, Wiley.
37. Ibid.
38. M. Johnson (1996) *Slaying the Dragon: How to turn your small steps to great feats*, Harper Collins. p. 9.
39. F. L. Smoll (2001) p. 153. Coach–parent relationships in youth sports: increasing harmony and minimizing hassle. In J. M. Williams, *Applied sport psychology: personal growth to peak performance*, Mayfield Publishing Company.
40. C. Brown, Boxing tests bond of father and fighter, The New York Times, July 15 2006. (http://www.nytimes.com/2006/07/15/sports/15boxing.html?pagewanted=print). Retrieved – 04/06/08.
41. R. Finn, A Life in the Fast Lane Takes Terrifying Turn, The New York Times, May 2, 1993. (http://query.nytimes.com/gst/fullpage.html?res=9F0CE0DB113AF931A35756C0A965958260). Retrieved – 04/06/08.
42. R. Pagliaro, Seles says good-bye, tennisweek.com, February 14 2008. (http://www.tennisweek.com/news/story_print.sps?inewsid=532725). Retrieved – 04/06/08.
43. Tributes to Sir Bobby Robson, The Daily Telegraph, November 16 2007. (http://www.telegraph.co.uk/sport/main.jhtml?xml=/sport/2007/11/16/sfnrob416.xml). Retrieved 04/06/08.
44. 'A leader of men is what he does best.' The Guardian. November 23 2004. (http://www.guardian.co.uk/football/2004/nov/23/newsstory.sport). Retrieved – 04/06/08.

Chapter 7

1. N. Comaneci (1981) *Nadia: My Own Story*, Proteus Publishing, p. 26.
2. Ibid.
3. Ibid.
4. S. Heady (1996) *Steffi: Public power, private pain*, Virgin Books.
5. Ibid., p. 11.
6. Ibid., p. 12.
7. I. Botham (2007) *Head On; The Autobiography*, Ebury Press, p. 7.
8. C. Giudice (2006) *Hands of stone; The life and legend of Roberto Duran*, Milo Books.
9. E. Woods (1997) *Training a tiger; the official book on how to be the best*, Hodder & Stoughton, p. xvi.
10. Ibid., p. xvii.
11. I. Botham (2007) *Head On; The Autobiography*, Ebury Press, p. 8.
12. J. McEnroe (2003) *Serious: The autobiography*, Time Warner Paperbacks p. 23.
13. Ibid.
14. M. Johnson (1996) *Slaying the dragon: How to turn your small steps to great feats*, Harper Collins.
15. B. Lorge, The tennis birthplace of the Deutschland duo, The SportStar, June 23 1990. (http://www.geocities.com/steffiarticles/90-02.htm). Retrieved – 04/06/08.
16. Ibid.
17. E. Woods (1997) *Training a tiger; the official book on how to be the best*, Hodder & Stoughton.
18. Ibid.
19. Ibid., p. 134.
20. Ibid., p. 131.
21. Ibid., p. 131.
22. Ibid., p. 149.
23. R. Philip (1993) *Agassi: The fall and rise of the enfant terrible of tennis*, p. 57.
24. Ibid.
25. E. Woods (1997) *Training a tiger; the official book on how to be the best*, Hodder & Stoughton, p. ix.
26. Ibid., p. ix.
27. J. Soutar (2006) *Ronaldinho: Football's flamboyant maestro*, Robson Books, p. 8.
28. I. Botham (2007) Head On: The Autobiography, p. 11.
29. Mike Tyson: Beyond the glory [documentary]. Fox Sports Net.
30. Ibid.

Chapter 8

1. C. Cheese, Murray rewarded for hard work, BBC.co.uk, 25 June 25 2005 (http://news.bbc.co.uk/sport1/hi/tennis/4618831.stm). Retrieved – 04/06/08.
2. S. Ahmed, Pro tour lures underclassmen, The Stanford Daily, February 23 2000. (http://daily.stanford.edu/article/2000/2/23/proTourLuresUnderclassmen). Retrieved – 04/06/08.
3. P. Tolme & M. Starr Kobe off the court, Newsweek.com, (http://www.newsweek.com/id/61773/page/3). Retrieved – 04/06/08.
4. M. E. Goodman (2003) Kobe Bryant, Creative Education, p. 15.
5. National Collegiate Athletic Association (2004) A career in professional athletics–A guide for making the transition, NCAA professional sports liaison committee.
6. Oscar De La Hoya: Beyond the Glory [documentary] Fox Sports Net.
7. Too much too young? BBC.co.uk,June 23 1999. (http://news.bbc.co.uk/1/hi/uk/376548.stm). Retrieved – 04/06/08.
8. Pete Sampras: Beyond the Glory [documentary] Fox Sports Net.
9. J. Morse Vaulting into Discord, Time Magazine, November 2 1998. (http://www.time.com/time/magazine/article/0,9171,989457,00.html?iid=chix-sphere) Retrieved – 04/06/08.
10. E. Pace, Christy Henrich, 22, Gymnast who suffered from anorexia, The New York Times, July 28 1994. (http://query.nytimes.com/gst/fullpage.html?res=9407E1DE1F3EF93BA15754C0A962958260). Retrieved – 04/06/08.
11. R. Seal Tales from the vaults, The Observer, December 4 2005. (http://observer.guardian.co.uk/osm/story/0,,1654132,00.html). Retrieved – 04/06/08.
12. Jones denies Balco claims, BBC.co.uk, April 25 2004. (http://news.bbc.co.uk/sport1/hi/athletics/3656721.stm). Retrieved – 04/06/08.
13. Disgraced Chambers in drugs claim, BBC.co.uk, May 28 2007. (http://news.bbc.co.uk/sport1/hi/athletics/6697417.stm). Retrieved – 04/06/08.
14. Powell Supports Chambers Claims, Eurosport.com, May 30 2007 (http://uk.eurosport.yahoo.com/30052007/4/powell-supports-chambers-claims.html). Retrieved – 04/06/08.
15. Ibid.
16. Quoted in P. David (2005) Human rights in youth sport, Routledge, p. 182.
17. P. Laure & C. Binsinger (2007) Doping prevalence among preadolescent athletes: a 4-year follow-up, *British Journal of Sports Medicine*, 41, 660–663.
18. W. Fotheringham Inquiry into Belgian cyclist's death raises new fears over EPO, The Guardian, February 16 2004. (http://www.guardian.co.uk/sport/2004/feb/16/cycling.cycling1). Retrieved – 04/06/08.

19. Flo Jo suffocated after fit, BBC.co.uk, October 23 1998. (http://news.bbc.co.uk/1/hi/world/americas/199306.stm). Retrieved – 04/06/08.
20. C. Henderson, Pantani death raises questions, BBC.co.uk, February 18 2004. (http://news.bbc.co.uk/sport1/hi/other_sports/cycling/3495967.stm). Retrieved – 04/06/08.
21. J. J. Coakley (1983) Leaving competitive sport: Retirement or rebirth? *Quest*, 35, 1–11.
22. A. Bodipo-Memba, Life after the NFL: Typically a struggle, USAtoday.com, January 28 2006. (http://www.usatoday.com/sports/football/super/2006-01-28-retirement-perils_x.htm). Retrieved – 04/06/08.
23. Top 10 lies that players believe, Gameover.org (http://www.gamesover.org/gonow/tenlies_players.cfm). Retrieved – 04/06/08.
24. T. Hauser, Mission: Impossible, Observer Sport Monthly, February 4 2007. (http://www.guardian.co.uk/sport/2007/feb/04/boxing.features). Retrieved – 04/06/08.
25. T. Graham, Holyfield has faith, but do we have it in him? ESPN.com (http://espn.go.com/boxing/columns/graham_tim/1385557.html). Retrieved – 04/06/08.
26. Boxing rallies round Bruno, BBC.co.uk, September 23 2003. (http://news.bbc.co.uk/sport1/hi/boxing/3131236.stm). Retrieved – 04/06/08.

index